A History of the Electron

Two landmarks in the history of physics are the discovery of the particulate nature of cathode rays (the electron) by J. J. Thomson in 1897, and the experimental demonstration by his son G. P. Thomson in 1927 that the electron exhibits the properties of a wave. Together, the Thomsons are two of the most significant figures in modern physics, both winning Nobel prizes for their work. This book presents the intellectual biographies of the father-and-son physicists, shedding new light on their combined understanding of the nature of electrons and, by extension, of the continuous nature of matter. It is the first text to explore J. J. Thomson's early and later work, as well as the role he played in G. P. Thomson's education as a physicist, and how he reacted to his son's discovery of electron diffraction. This fresh perspective will interest academic and graduate students working in the history of early twentieth-century physics.

JAUME NAVARRO is Ikerbasque Research Professor at Universidad del País Vasco/Euskal Herriko Unibertsitatea. He trained in physics, philosophy and the history of science, and has an international research record, having spent several years at the University of Cambridge, Imperial College London, the Max Planck Institute for the History of Science and the University of Exeter.

A History of the
Electron

J. J. and G. P. Thomson

JAUME NAVARRO

Ikerbasque Research Professor,
Universidad del País Vasco/
Euskal Herriko Unibertsitatea

CAMBRIDGE
UNIVERSITY PRESS

University Printing House, Cambridge CB2 8BS, United Kingdom

One Liberty Plaza, 20th Floor, New York, NY 10006, USA

477 Williamstown Road, Port Melbourne, VIC 3207, Australia

314-321, 3rd Floor, Plot 3, Splendor Forum, Jasola District Centre, New Delhi - 110025, India

79 Anson Road, #06-04/06, Singapore 079906

Cambridge University Press is part of the University of Cambridge.

It furthers the University's mission by disseminating knowledge in the pursuit of education, learning and research at the highest international levels of excellence.

www.cambridge.org
Information on this title: www.cambridge.org/9781108724432

First published 2012
First paperback edition 2019

A catalogue record for this publication is available from the British Library

Library of Congress Cataloging in Publication data
Navarro, Jaume.
  A history of the electron : J. J. and G. P. Thomson / Jaume Navarro.
    pages   cm
  Includes bibliographical references and index.
  ISBN 978-1-107-00522-8
  1. Electrons–History.   2. Cathode rays.   3. Thomson, J. J. (Joseph John),
  1856–1940.   4. Thomson, G. P. (George Paget), 1892–1975.   I. Title.
  QC793.5.E62N38   2012
  539.7´211–dc23   2012018834

ISBN 978-1-107-00522-8 Hardback
ISBN 978-1-108-72443-2 Paperback

To my parents, Rupert and Maria Teresa

Contents

Introduction

In 1897, Joseph John Thomson, Professor of Experimental Physics and director of the Cavendish Laboratory in Cambridge, ascribed corpuscular nature to the carriers of electricity in cathode rays. This event constitutes a central element in what is traditionally known as the *discovery* of the electron. Exactly 30 years later, his son, George Paget Thomson, obtained the first ever images of electron diffraction, with which he showed the wave-like behaviour of his father's electrons. Ironically, while the father had shown that a wave phenomenon (cathode rays) could be explained in terms of corpuscular entities (electrons), the son was reclaiming wave characteristics for his father's corpuscles. This is, in a nutshell, the story of this book, one that, however familiar to many physicists and historians of science, has never been told in detail.

Alongside this father-and-son narrative, I intend to explore a number of historiographical and philosophical questions. To begin with, this book is biographical, but not a traditional biography. The main characters on the stage are J. J. Thomson, G. P. Thomson and the corpuscle-electron, and the book deals with the intersection of their lives. There are a number of partial biographies of J. J. Thomson, starting with his own memoirs (Thomson, 1936; Rayleigh, 1942; G. P. Thomson, 1964; Davis and Falconer, 1997; Kim, 2002), but nobody has so far written a complete biography of him. Just from counting the pages in this volume, it is clear that this book is no attempt to plug this bibliographical lacuna. But it does enter into an aspect of his career often overlooked: his science in the 1920s, when a young generation of physicists regarded him as a relic of times past. As I shall argue in Chapters 5 and 6, an examination of his work in those years sheds some new light on the content and motivation of the better-known aspects of his career. Nor does a biography of G. P. Thomson exist. Indeed, there has been surprisingly little historical work done on him for a man who was inter alia a Nobel laureate in physics, chairman of the MAUD committee, and Master of Corpus Christi College, Cambridge.

As for the electron, a number of *biographies* of this entity do exist, from Charles Gibson's (1911) *Autobiography of an Electron* to Theodore Arabatzis' (2006) *Representing Electrons*, the latter being a very good case study for a biography of an epistemic object (Daston, 2000). The present book treats the life of the electron only insofar as it impinges on the Thomson household, from its conception as a corpuscle for explaining cathode rays and the conduction of electricity in gases to its maturity as the first quantum particle with wave properties. We shall see that, as an epistemic object, this electron underwent a number of upheavals and personality changes (continuing with the metaphor) in order to play an explanatory role in an increasing number of phenomena: from cathode rays and atomic constitution to electronic radiation, chemical bonding and quantum indeterminacy. We shall see how this flexibility was possible partly on account of its ontological under-determination. In other words, J. J Thomson's electron did not exist as an ultimate explanation of what matter *is*, but rather as an epistemic object mediating between observed phenomena and the ultimate reality of a continuous ether and the Faraday tubes therein.

As a matter of fact, one could read this book as a biography of Faraday tubes rather than the electron. To be sure, the ether itself is the underlying entity in this story, since it was the ether that embodied J. J. Thomson's belief in a metaphysics of the continuum. Faraday tubes began as a concrete mental model within this continuous framework and slowly almost acquired physical reality in his theoretical work. As we shall see, Thomson expected them to be the underlying entity that might explain the nature of electrons and the seemingly contradictory properties of different forms of radiation. Soon after J. J. Thomson had found evidence for the atomicity of electric charge, physics began to move towards a more fundamental atomicity, that of energy. With the new quantum theory, the fundamental tenets of Thomson's world-view began to tremble, the existence of the ether was in jeopardy, and the laws of classical physics proved to be insufficient for explaining certain phenomena. The problem of accounting for the apparently self-contradictory behaviour of radiation, at times like corpuscles, at other times like waves, loomed large in Thomson's work, which famously led him to coin the expression that physicists were witnessing *a struggle between a tiger and a shark*. Faraday tubes were, for him, the tool that might educe harmony between continuity and discontinuity in the physical world.

Furthermore, G. P. Thomson's experimental demonstration of the wave-like behaviour of electrons confirmed J. J.'s lifelong-held metaphysical views that gave priority to a continuous medium (the ether) rather than to the corpuscular explanations of matter (the electron), the latter being, in his own words, *a policy, not a creed*, on the nature of matter. That is why both father and son could easily accept electron diffraction, since it did not seriously challenge, but rather somehow confirm, their classical world-view. In a way, the Thomsons saw in one of the tenets of the new quantum (i.e., discrete) mechanics, wave–particle duality, confirmation of the classical notion of continuity.

J. J. Thomson's discovery of the electron has always been historiographically problematic. As Isobel Falconer (1985, 1987, 1989) thoroughly showed, his research, from his appointment as Professor of Experimental Physics in 1884, had little to do with cathode rays. He was mainly interested in electric discharges in tubes as a way of understanding the relationship between matter, electricity, and chemical bonding. His interest in cathode rays only came about after the discovery of X-rays in 1895. His electron – the corpuscle – was gradually introduced into explanations of the conduction of electricity, the composition of matter, and chemical bonding. The different uses of the electron were crucial in determining its ontological status: first, in his attempts to obtain a complete explanation for the conduction of electricity and the interaction between electricity and matter, later (and only later) in its role as a subatomic particle. The notion of the corpuscle also permeated his project on positive rays, which he started around 1907, and in which he tried to emulate his experiments on cathode rays in the search for some possible *corpuscle* of positive electricity. He also made frequent incursions into the territories of chemistry, his 1923 book *The Electron in Chemistry* being the highlight of these attempts. The broader picture of the uses of the electron in the first decades of the twentieth century can be found in Buchwald & Warwick (2001).

This biographical sketch of the Thomsons, father and son, of the electron, and of the ether and its Faraday tubes is set on a very particular stage: the University of Cambridge and the Cavendish Laboratory. The book follows in the footsteps of Andrew Warwick's (2003a) *Masters of Theory*, wherein he uses the pedagogical traditions of the Mathematical Tripos as the explanatory tool to evaluate the strengths and weaknesses of the theoretical physics that was done in that university at the end of the nineteenth century and beginning of the twentieth. In that book, Warwick justified the particular way Cambridge (and, by extension, British) physicists understood the new theory of relativity as the almost inevitable outcome of local pedagogical régimes. Here, we shall find a somewhat analogous story in the case of the early theory of the quantum. J. J. Thomson's reaction to the emergence of the new theory, one that challenged the very essence of what science was for him, may be easily interpreted as paradigmatic of a generation of scientist for whom physical explanations had to involve some kind of mental model. Less obvious is G. P. Thomson's case. Trained in Cambridge in the years immediately before the Great War, he was a contemporary of Niels Bohr and could have, in principle, played an active role in the reception and development of quantum physics in Britain, more so when we consider how his experiments on electron diffraction constituted proof of one of the most counterintuitive principles of the new physics. But, as we shall see, he did not construe his work as an *experimentum crucis* standing between the old and the new physics, but merely the confirmation that an explanation in corpuscular/discrete terms only was not enough to resolve the problems of radiation and account for the identity of subatomic particles.

Cambridge will also figure in an unexpected way. After his graduation, G. P. Thomson joined his father in pursuing a research project on the nature of positive rays, one that began as an attempt to understand the nature of positive electrification, later to evolve into a technique of chemical analysis. After the Great War, during which father and son each worked on different military-related projects, they went back to working together again. I shall refrain from making psychological speculations on the kind of emotional dependency the son had on the father, for which we do not have specific evidence. What is of interest is that this reunion happened at a time when J. J. Thomson had been encouraged to step down as director of the Cavendish Laboratory after 35 years' tenure, and that the collaboration did not cease even when, in 1922, G. P. Thomson was appointed as a professor at the University of Aberdeen. As I shall argue in Chapter 6, the son used this first opportunity he had of running a research laboratory, however small it was, to replicate in Aberdeen the experimental set-up he had with his father in Cambridge, virtually turning his laboratory into an extension of his father's old rooms at the Cavendish.

There is one last parallel between the father's and the son's experimental work on electrons that I highlight in this book. As we shall see in Chapter 3, J. J. Thomson had been working on electrical discharges in tubes filled with gases for over ten years when X-rays appeared on the stage. The latter were obtained in cathode-ray tubes, i.e., evacuated tubes, and Thomson's experimental setting could rather easily be modified to analyse the new X-rays and the old cathode rays. His famous experiments of 1897 were, in a way, the serendipitous outcome of a long project that had positively avoided the use of cathode rays. Similarly, in 1923, Louis de Broglie suggested that electrons, and indeed any other particle, could be understood as possessing a dual nature: wave and particle. In the summer of 1926, many British physicists learnt of de Broglie's theory and Schrödinger's developments in quantum mechanics, and it so happened that G. P. Thomson was in a privileged situation to put this principle to the test. His experimental set-up in Aberdeen for studying positive rays could, again, rather easily be adapted to take pictures of the possible diffraction of cathode rays through thin metallic films. Both the discovery of the electron by the father and that of its diffraction by the son were possible thanks to the quick modification of experimental set-ups originally meant for other projects.

Whether matter and radiation are by nature continuous or discrete are central considerations in this book. But I shall also discuss the continuity, or lack thereof, between scientific disciplines, specifically between physics and chemistry, and the nascent field of physical chemistry. The emergence of the last is largely related to the explanatory power of electrons for chemical bonds. In Chapter 2, we shall find, however, that, from the early 1880s, J. J. Thomson had in mind the idea of uniting physics and chemistry under the common umbrella of *the physical sciences*, on both the conceptual and the institutional levels. That particular project did not succeed, but it revises the customary image of J. J. Thomson, who can be understood better as a practitioner of the physical sciences than simply as a physicist. This will also become evident in Chapter 5, when we find the Thomsons' project on positive rays developing into an experimental method of chemical analysis.

Now it is time for acknowledgements and expressions of gratitude. Lest the reader be bored, I shall only mention those who had a very active role in the genesis of this book, starting with Andrew Warwick, who has lent much-needed support at various stages of this project. I should also like to mention Simon Schaffer and Richard Noakes, in Cambridge, and Massimiliano Badino and Shaul Katzir, of the Max Planck Institute for the History of Science, for their insightful ideas and their very helpful advice, as well as the ideas I received from conversations with Hasok Chang and Isobel Falconer, among others. I also want to thank Sebastian Hew for his patient and thoughtful editing of my writing. I am indebted to staff at the following archives for their help and for granting permission to work and cite from their materials: Cambridge University Library (CUL), Royal Society (RSA), Royal Institution of Great Britain (RI), University of Aberdeen (UAb), Trinity College and Churchill College. Lastly, I want to thank Mr David Thomson, son of G. P. Thomson and grandson of J. J. Thomson, for his time and help, and for permission to quote from his father's archives and autobiography in Trinity College, Cambridge.

1

The early years in Manchester and Cambridge

1.1 Manchester

Chimneys are the main architectural element that characterize our first destination, Manchester in the mid-nineteenth century. Chimneys have dwarfed the bell towers of the old provincial town, and the clangour of the looms has silenced the bells. The power of the steam engine, of the free market, and of enterprise has transformed the city into the centre of what we now call the Industrial Revolution: a landscape of chimneys, like those later portrayed in the paintings of L. S. Lowry, ceaselessly belching smoke into the always-humid air of the Lancashire region. As a contemporary observer put it, 'the clouds of smoke vomited forth from the numberless chimneys, Labour presents a mysterious activity, somewhat akin to the subterranean action of a volcano' (Fraucher, 1844, p. 2).

Smoke from the factories mingled with steam, with clouds, and with fog, forming all sorts of capricious combinations of fluids and giving rise to playful shapes in the atmosphere. In the mid-nineteenth century, the citizen of Manchester was constantly breathing the insalubrious air created by the industrial machinery: an air that imbued everything in the city, impregnating clothes, buildings and the deep corners of every lung, with terrible odours, dirt and all sorts of diseases. Not only did the Manchester air, like that other entity of Victorian science, ether, permeate many aspects of life, irrespective of social class or age, it was also the source of awe for those interested in the study of fluids, their mixtures, their shapes and forms, and their diffusion through solid bodies. The skies in Manchester became a privileged environment wherein to observe the behaviour of smoke rings, diffusion patterns, and condensation phenomena, all of which were part of the interest of the Victorian men of science.

The better-off classes were also fortunate to experience the different concentrations of that air, since they tended to live in the suburbs of the city, where the atmosphere was significantly cleaner. In a most poignant description made in 1844, Friedrich Engels pointed out the fact that 'Outside, beyond this girdle, lives the upper and middle bourgeoisie, the middle bourgeoisie in regularly laid out streets in the vicinity of the working quarters … the upper bourgeoisie in remoter villas with gardens in Chorlton and Ardwick, or on the breezy heights of Cheetham Hill, Broughton, and Pendleton, in free, wholesome country air, in fine, comfortable homes' (Engels, 1845/1887).

The ubiquitous chimneys were only the tip of the iceberg of the changes that Manchester underwent in the early nineteenth century. In less than a hundred years, the city became the region with the highest density of population in England, seeing a ten-fold increase in the number of inhabitants. The following figures speak for themselves: the population grew from about 24 000 in 1773 to over 300 000 in 1841. The figures, however, belie the most significant change in the structure of the population. At the end of the eighteenth century, in the Mancunian population there was a provincial elite of clergymen, physicians and small-scale traders. By the mid-nineteenth century a growing bourgeoisie of textile industrialists had replaced this elite.

Besides the old and new elites, thousands of working-class people were crammed into neighbourhoods built specifically for them where they lived in subhuman conditions. The descriptions of Friedrich Engels, however exaggerated they may be, give us a vivid account of the landscape: 'Of the irregular cramming together of dwellings in ways which defy all rational plan, of the tangle in which they are crowded literally one upon the other, it is impossible to convey an idea … the confusion has only recently reached its height when every scrap of space left by the old way of building has been filled up and patched over until not a foot of land is left to be further occupied'. And with this chaotic and very dense concentration of human beings came the highest degree of filth and insalubrious conditions: 'In dry weather, a long string of the most disgusting, blackish-green, slime pools are left standing on this bank, from the depths of which bubbles of miasmatic gas constantly arise and give forth a stench unendurable even on the bridge forty or fifty feet above the surface of the stream' (Engels 1845/1887). The social structure of Manchester and its division of labour was also embodied in the strict separation between the different neighbourhoods:

> The town itself is peculiarly built, so that a person may live in it for years, and go in and out daily without coming into contact with a working-people's quarter or even with workers, that is, so long as

he confines himself to his business or to pleasure walks. This arises chiefly from the fact, that by unconscious tacit agreement, as well as with outspoken conscious determination, the working-people's quarters are sharply separated from the sections of the city reserved for the middle-class; or, if this does not succeed, they are concealed with the cloak of charity ... And the finest part of the arrangement is this, that the members of this money aristocracy can take the shortest road through the middle of all the labouring districts to their places of business without ever seeing that they are in the midst of the grimy misery that lurks to the right and the left.

This social structure has also strong parallelisms with the main entity of Victorian physics: the ether. If the mixture of smoke, fog, and air created a privileged image for the structure of the ether, the social structure of Manchester resembled the relationship between ether and the material world. The working class, together with the coal, would have been the invisible power behind the rise in production. The consumer, the middle and upper classes, would only see the result of the process, without getting into the minutiae of the conditions of the working class, their activities, and their work. Analogously, the invisible ether would permeate the activities of the visible world, being the see for diverse forms of energy and, in some cases, also the see for the spiritual world. The ether permeated the whole of the cosmos in Victorian science, and will be present throughout this book. That is the reason for starting with these two analogies that can be found in mid-nineteenth-century Manchester, the hometown of Joseph John Thomson.

1.2   Science in Manchester

The profound changes in the social structure of Manchester triggered in its citizens a transformation in their approach to science. At the end of the eighteenth century, science was almost non-existent in Manchester. Located between the scholarly centres of the south (Oxford, Cambridge and London) and the north (Scotland), Manchester was something of an academic desert. The Mancunian gentry were content with the less-than-exciting intellectual life of local institutions, such as the parish halls, the libraries, the clubs, and the amateur theatres. In just over 50 years, however, the panorama had changed completely. Triggered by the new economic situation, science developed in Manchester in the first decades of the nineteenth century basically because of two factors. On the one hand, the development of industrial machinery and technologies stimulated an army of engineers, chemists and other technical personnel, creating the need for spaces to exchange information. On the other hand, some of the entrepreneurial traders and industrialists that populated the city felt the need to relate to nature in a purer way than industry allowed. This gave rise to a particular brand of *savants of nature* composed of successful industrialists for whom the social status came from their philosophical interest in nature, not from the money they made using its resources.

At the turn of the nineteenth century, the only institution in which natural philosophy was somewhat present was the Manchester Literary and Philosophical Society (the Lit & Phil), which originated in the last third of the eighteenth century as informal meetings of mainly medical doctors and was formally established in 1781 (Cardwell, 2003). Initially, the number of ordinary members could not exceed 50, and these were elected on the grounds of their residency in Manchester and surrounding areas, and most importantly, on the basis of their literary or philosophical contributions. The meetings of the society, and the subsequent *Memoirs*, dealt with topics such as natural philosophy, chemistry, literature, civil law, commerce and the arts. Excluded from these debates were British politics, religion, and the practice of medicine in an attempt to avoid belligerent disputes among the members of the Society. In due course, other scientific institutions appeared in Manchester: in 1821, the Natural History Society was established, and soon after the Royal Manchester Institution; in 1825, the Manchester Mechanics' Institution; in 1829, the New Mechanics' Institution; in 1839, the Salford Mechanics' Institution; in 1834, the Statistics Society, and in 1838 the Manchester Geological Society. However, the Lit & Phil maintained pre-eminence over the rest.

Most members of the Lit & Phil were amateur intellectuals. They had their jobs as doctors, chemists, industrialists or tradesmen, but devoted part of their time to the gentle cultivation of the sciences or the arts. Among this very amateur and dilettante tradition, and in contrast with it, we find the best-known Mancunian natural philosopher, John Dalton, a self-trained natural philosopher, who became a member of the Society in 1794 and presided over it from 1817 until his death in 1844.

Dalton's background and attitude towards science meant a first turning point in the nature of the Lit & Phil. He was not a well-established professional, nor did he come from a bourgeois background. He was born into a family of religious dissenters, for whom many educational institutions were, at the time, banned. He started his career in the context of Quaker educational institutions, and arrived in Manchester as a teacher of natural philosophy in a newly created college. His work on meteorology soon gave him local prestige, a prestige he used to become a full-time man of science. The members of the Lit & Phil agreed to let Dalton work in rooms of the Society, which he equipped at his own expense, and from which he emerged as an internationally renowned natural philosopher. It was in the setting of the Lit & Phil that Dalton developed his atomic theory of matter, giving precise, quantitative data on the proportion of the different elements in chemical compounds. Dalton's name would, for evermore, be linked to the atomic theory of matter.

The importance of Dalton as an icon of Manchester science was particularly clear during his funeral, on 12 August 1844, an event that was tailored by the local authorities to signal Manchester as a place for first-rate natural philosophy. According to the reporter in *The Manchester Guardian*, the chapel of rest, installed in the Town Hall, was visited in one day 'by no less than forty thousand people'. In the procession to the cemetery, 'nothing could be more gratifying than the quiet, orderly behaviour, and the silent and respectful demeanour, of the immense concourse of persons along the whole distance ... The shops were closed; ladies and gentlemen, in mourning, filled every window ... Indeed, we never saw in this community so general a wearing of mourning attire, crape, &c.' (*The Manchester Guardian*, Wednesday 14 August, 1844). Such ostentatious display was condemned only by the Society of Friends, to which Dalton had belonged.

Dalton was succeeded by James Prescott Joule as the icon of Mancunian science. Born in 1818, Joule became, by the 1850s, its most visible face. The son of a very successful Salford brewer, he was trained by Dalton in the mid-1830s and did most of his science in the laboratory that he set up in his home. There he developed the ideas and experiments that would eventually lead him to formulate his ideas on the transformation of different forms of energy, including the paddle-wheel experiments for which he became known. 'When I was a boy', Thomson recalled, 'I was introduced by my father to Joule, and when he had gone my father said, "Some day you will be proud to be able to say you have met that gentleman"; and I am' (Thomson, 1936, p. 10).

Joule spent every day from nine to six in the brewery, trading and dealing with his father's business. But his true calling was in his home laboratory, among the electrical and chemical apparatus with which he was experimenting. As soon as he could after his father's death, he sold the brewery and fully embraced his passion for science. Joule's shift from amateur to professional science is paradigmatic of Manchester in the mid-nineteenth century. Two elements played a major role in the change of attitude towards science: first, industry was growing, and its needs were more and more sophisticated. They demanded a more professional approach, far from the idealism and seclusion of men like Dalton. Second, the new bourgeoisie began to feel science was a calling, and not an elegant pastime, and thought they had to become savants of science, as if industrial activity was not *pure* enough. To follow their calling in science, they started as self-trained scientists (many were self-made businessmen). Their entrepreneurial drive included the research and publication of cutting-edge scientific knowledge. In the words of historian R. H. Kargon, 'this new group, generally from the less prestigious segments of the middle class and sometimes self-made men, were *devotees* of science, who saw science as their "calling" ... These businessmen-savants, however, sought their *identity* and ... even their *status* in the scientific pursuit' (Kargon, 1979, p. 35).

Under Joule's influence, the meetings and the *Memoirs* of the Lit & Phil also underwent dramatic changes in their scope and content. They became more and more the forum for almost exclusively scientific papers. While, at the beginning of the century, only 50% of papers in the *Memoirs* were related to the sciences, by the middle of the century, this proportion was 95%, including engineering, which increasingly occupied an important place in the meetings of the Society. The revolution that was taking place in France and Germany, where industry was increasingly collaborating with scientific institutions, was also, although a bit late, coming to Manchester. Joule managed to change the regulations of the *Memoirs* of the Lit & Phil so that it became a scientific journal *tout court*, including the dates of the reception of manuscripts, and de facto excluding all literary articles. In due course, the naturalist *devotee* was to give way to a totally professional practitioner of science. Unlike their predecessors, however, this new group of scientists became established in Manchester after they were trained in other British and European research centres. Their primary task was to overcome the remnants of amateurism that still pervaded in Manchester and to establish and consolidate scientific institutions in the city.

Many devotees also saw in their scientific endeavours a somewhat religious mission. As an example, we can reproduce the notes that Joule prepared for his speech to the British Association for the Advancement of Science meeting of 1873:

> The great object which natural science has in view is to elevate man in the scale of intellectual creatures by the exercise of the highest faculties of his nature in developing the wonders of the glorious creation. The second and subsidiary object is to promote the well being and comfort of mankind to increase his luxuries. These objects are closely allied and should not be separated. The benefit to be attained is for the entire man, for his soul, his mind, his body. The importance of this object is measured by the importance of that part of human nature which is beneficially affected. The first object is therefore at least as much more important than the second as the intellect is more noble than the body ... And yet it is evident that an

acquaintance with nature's law means no less than an acquaintance with the mind of God therein expressed. This acquaintance brings us nearer to him... (Kargon, 1979, pp. 55–6).

Dalton and Joule were the two best-known physicists of nineteenth-century Manchester. The first became internationally renowned for his work in support of the existence of atoms as corpuscles of matter; the latter became the icon of the conservation and transformation of different forms of energy. Both concepts eventually became central to the scientific career of J. J. Thomson, who became known for the discovery of the corpuscle-electron while retaining a metaphysical framework in which the conservation of energy in the ether played a crucial role. At the very beginning of his memoirs Thomson, a proud Mancunian, pointed at the relevance of these two names in the configuration of science in Manchester: 'Manchester has played a prominent part in the history of physical science, for, in it, in the first half of the nineteenth century, Dalton made the experiments which led him to the discovery of the law of multiple proportion in chemical combination, and Joule those which were instrumental in establishing the principle of the Conservation of Energy' (Thomson, 1936, p. 7). J. J. was to think of himself as the one who was able to unify these concepts.

1.3   Thomson's early days

Joseph John Thomson was born in December 1856 in Cheetham, one of those Manchester suburbs accommodating the local middle-class bourgeoisie of which Engels spoke in his description of the city. His father, Joseph James, ran a modest publishing and bookselling business that three generations earlier one Ebenezer Thomson had founded in Manchester. His mother's name was Emma Swindells. The family would be completed two years later with the birth of J. J.'s only brother, Frederick Vernon.

Little is known about J. J.'s life as a child. The only information we have are the recollections he recorded in his autobiography, written around the age of 80. However, a few points are relevant for the present story. Thomson was born into a middle-class family, and educated in a local school. 'After going for a year or two to a small school for young boys and girls, kept by two maiden ladies who were friends of my mother', he wrote, 'I went to a private day school kept by two brothers named Townsend, at Alms Hill, Cheetham, which was near to where I lived' (Thomson, 1936, p. 2). In the British context, this needs to be emphasized, since it was not uncommon for the offspring of the establishment to be educated, from an early age, in the prestigious boarding public schools scattered all over the country. Education in that kind of establishments was a good starting point from which to gain access to the traditional universities of the country. Thomson was not, therefore, on the right track from which to start an academic career.

As we shall see in Chapter 5, later in life, J. J. Thomson became very active in educational policies, with very strong views on the way children should be taught. He looked back with a certain sense of nostalgia on some of the aspects of his earlier education, especially the promotion of the use of memory. He was glad to have studied Latin following the Eton Latin Grammar, which was written in verse, so that it was easier to memorize, even though the meaning of words was not always clear. In English, he had to learn by heart fragments of the works of keynote authors such as Shakespeare, Byron, and Scott. The syllabus also included some history and a great deal of arithmetic, which 'is an excellent intellectual gymnastic, for it is easy to set simple questions which cannot be solved by rule of thumb, but require thought' (Thomson, 1936, p. 4). There was almost nothing of the natural sciences at school, except for the collection of some animal and botanical species. Thomson, however, from a very early age cultivated the Victorian hobby of growing rare species of flowers, 'and thought that when I grew up I should like to be a botanist' (Thomson, 1936, p. 6). His son confirmed that, besides physics, his father's other passion in Cambridge was his garden (G. P. Thomson, 1966, p. 55). We also know that he possessed a small microscope his father had given him as a present.

Life in the Thomson household seems to have been pleasant, with the normal comforts of a Victorian middle class family. Because of his occupation, Thomson's father was relatively well connected to the intellectual elite of the city, which gave the young J. J. the opportunity to meet some local great men like James Joule. Everything changed, however, with the death of his father when J. J. was just 16. Although they received help from many friends, the family had to move to a smaller house and their straightened circumstances meant that J. J.'s younger brother was unable to receive the same education as J. J.

Thomson seems to have been very fond of his brother Fred, who took on the responsibility of looking after their mother when J. J. left for Cambridge. Fred worked in a local calico merchant company with strong connections in the USA and lived with his mother until her death in 1902, normally meeting with J. J. in the summer vacation. In 1914, he fell seriously ill and J. J. suggested that he should move to Cambridge, so that he and his wife could look after him. Fred rented a house near his brother's until his death three years later, in 1917. The affection of J. J. for his brother was very obvious, and he 'always personally carried a wreath to his grave at Christmas for 21 years, after which he was no longer physically able to do so' (Rayleigh, 1942, p. 185).

1.4   Owens College

In spite of his early passion for botany, Thomson, like so many young people in Manchester, intended to become an engineer. At the time, there was no such thing as an academic career in engineering in Britain, and the only way to become one was through hands-on training as an apprentice in an engineering company. However, due to the existence of a long waiting list for such positions, his father decided to send him to Owens College at the age of 13. There, he proved his abilities in pure science, which drew the attention of one of his teachers, Thomas Baker, himself a former Cambridge graduate, who eventually encouraged him to sit for a fellowship at Trinity College. Actually, Thomson was not the first Owens student to go to Cambridge. The year he started at Owens, John Hopkinson, who had also studied at the college, graduated as senior *wrangler* – the highest honour – in Cambridge, which gave Owens College much prestige. This seems to have been a strong argument to convince people such as Mr Thomson to send their children to Owens.

J. J. was also grateful for the opportunity to go to Owens College, and saw this 'accident' as 'the most critical event in my life and which determined my career' (Thomson, 1936, p. 2), since the college was a unique institution, unparalleled in Britain, both for the scope of its education and for the young age at which students could be admitted. The curriculum at Owens was very different from that of most British schools, since it stressed the importance of engineering, experimental physics and chemistry and put less emphasis on the humanities and mathematical physics (Kargon, 1979, pp. 157–96). The goal of the junior school which J. J. attended was to prepare young students for the formal study of the natural sciences, with an eye to training scientists who would, directly or indirectly, make a difference to the development of Manchester's scientific and industrial network.

What kind of institution was Owens College? From the 1830s, a number of influential people advocated the idea of setting up a university in the industrial and commercial city of Manchester as the necessary culmination of the great development of the metropolis. However, it only materialized after the death of John Owens, a wealthy Manchester merchant whose will included a large sum of money to be used for the establishment of a university in Manchester. In 1851, Owens College became a reality, as a college affiliated to the University of London. In the beginning, three chairs were created: one for classics, one for mathematics and natural philosophy, and a third one for mental and moral philosophy and English, following the patterns of traditional universities. There were also part-time appointments in chemistry, in botany, zoology and botany (natural history), in German and in French.

Although the chemistry appointment was only part-time initially, chemistry eventually evolved into one of the key areas at Owens College, partly due to the tradition started by Dalton and continued by Joule, but also partly due to the technological needs of the many industries in the region. Nevertheless, the establishment of an academically serious department of chemistry was not without obstacles. The first professor, Edward Frankland, had trained with two of the major figures in mid-nineteenth century chemistry (in Marburg, under Robert W. Bunsen, and in Giessen under Justus Liebig), and he had great expectations of replicating something like Liebig and Bunsen's laboratories in Manchester. Although, at first, his was only a part-time appointment, Frankland managed to organize a chemistry laboratory and wanted to emphasize the importance of his discipline, not only for practical industrial and medical reasons (which suited potential Manchester students very well) but also for the pursuit of knowledge per se. He tried to 'recommend the science [of chemistry] for its own intrinsic excellence, for the intellectual delight which every student must find in its pursuit and for the bright glimpses of the Deity which it discloses at every step' (Kargon, 1979, p. 159). But Frankland did not succeed in establishing a research department, partly because the few students he had seemed to be interested only in 'testing the "Soda-ash" and "Bleaching Powder"'(Frankland to Bunsen, cited in Kargon (1979 p. 164), and, tired and disappointed, he decided to leave the college in 1857.

Henry Enfield Roscoe was Frankland's successor in the chair of chemistry. He had also trained with Bunsen and Liebig in Germany, where he learnt of the benefits of research laboratories with links to commercial enterprises. He succeeded in using Frankland's initial impetus to build, over the course of ten years, a school of chemistry with higher quality theoretical and experimental training, industrial links, and original research. A second chair in chemistry was created in 1874, and Carl Shorlemmer, Roscoe's assistant, born and trained in Germany, was appointed, turning chemistry into a central science at Owens. In his *Recollections* of 1936, Thomson attributed the success of Owens to Roscoe's interest in promoting education in the sciences, acknowledging that 'By the time he left Owens, its Chemical Department had become the best organized and the best equipped in the country, attended by over a hundred students'. A main reason for this success was that 'Roscoe did much by his own personal efforts to promote the application of science to industry. The manufacturers in Lancashire believed in him and were constantly coming to consult him as to the way they should get over difficulties which had cropped up in their work' (Thomson, 1936, p. 29). Roscoe had also become interested, while in Germany, in the new science of spectroscopy, a field that proved very successful in the last decades of the nineteenth century. His book on the analysis of spectra turned out to be quite influential in the laboratories engaged in spectroscopic techniques, one of which from the late 1870s onwards was, as we shall see in Chapter 2, the Department of Chemistry of the University of Cambridge.

But let us go back to 1857. Only six years after the opening of the college, the decline in the number of students proved that more encouragement was needed to attract the middle classes into higher education. In Manchester, the offspring of the emergent middle class were mainly interested in practical training and short-term profit, and few people enrolled in fundamental research. Only slowly did the ruling class in Manchester, and the authorities at Owens, realize that the college could neither aspire to compete with the traditional liberal education of Oxford and Cambridge, nor content itself with providing superficial vocational training. The French, and especially the German, scientific institutions were improving their techniques and productivity at a faster pace than the British, thanks to a radically new way of merging academic research and industrial interests, and this was the pattern that Owens College tried to follow after 1857 (Sviedrys, 1976; Gooday, 1990; Turner, 1993). Also in that year, it was decided to start a junior school associated with the college, the main purpose of which was to get people interested in fundamental research from a younger age. This was the school in which J. J. Thomson enrolled in 1870. In 1873, Thomson saw his college move into new buildings on Oxford Street, whose neo-gothic architecture expressed the merging of industry, commerce, training and research.

The academic staff that J. J. found in the junior college were extraordinary for a teenager engineer-to-be: in the chair in mathematics, Thomas Baker, a senior wrangler and Fellow of Trinity College, Cambridge; in engineering, Osborne Reynolds, who had also been trained in the Cambridge Mathematical Tripos after an apprenticeship in mechanical engineering; in physics, Balfour Stewart, who came from the Scottish universities of St. Andrews and Edinburgh; and in chemistry, the aforementioned Roscoe and Schorlemmer. In his autobiography, Thomson recalled the influences that he received from all of them.

From Baker, he received his first instruction in advanced mathematics, which enabled him to learn about the newly introduced quaternionic notation long before he attended Cambridge. Thus, from his early days, Thomson was accustomed to using powerful analytical tools in the solution of physical problems, a tradition that he would certainly develop in the Cambridge Mathematical Tripos. Reynolds was a seventh wrangler, but had had a four-year apprenticeship in engineering before going to Cambridge. He was appointed to the newly created chair of engineering at Owens in 1868. Thomson described him in his autobiography as 'one of the most original and independent of men, and never did anything or expressed himself like anybody else' (Thomson, 1936, p. 15).

Taking notes during his lectures proved to be almost impossible, and the students had to rely on the well-known textbooks by the Scottish physicist and engineer William J. N. Rankine's textbooks; but it was also evident to them that Reynolds had a very original and independent mind. Since Thomson was initially waiting to be admitted to an apprenticeship in engineering, he spent a lot of time with him in the early years.

But as some historians have emphasized, it was Balfour Stewart who probably most influenced Thomson at Owens College (Davis & Falconer, 1997, p. 6; see also Crowther, 1974; Chayut, 1991). On the one hand, he introduced J. J. to Maxwell's recent *Treatise of Electricity and Magnetism* (Maxwell, 1873), arousing his interest in this science. On the other hand, Stewart was passionate about teaching in the laboratory, and he introduced Thomson into hands-on practical research. They spent long hours together, engaged in laboratory work, trying, for instance, to detect a change of weight in chemical reactions, with even an accident taking place in which Thomson apparently nearly lost his sight. Although Stewart was formally a professor of physics, his research topics were at the boundary between physics and chemistry, which contributed to Thomson's idea that both disciplines were part of a bigger whole. For example, Stewart organized practical courses for three kinds of students, including those 'who wish to confine themselves to those branches of Physics most allied to Chemistry' (Chayut, 1991, p. 532). In his study of Thomson's inclination towards chemistry, Sinclair emphasized that Stewart's ideas on the conservation of energy, the constitution of matter, the nature of the ether and of the atoms, etc. were extremely influential on J. J., which explains why he always saw many physical problems 'in something of a chemical light' (Sinclair, 1987, p. 91).

J. J.'s time in this institution did not pass unnoticed. His abilities in the sciences gained him several local prizes and scholarships, such as the Ashbury Engineering Scholarship, the Dalton Junior and Senior Mathematical Scholarships and the Engineering Essay Prize (Rayleigh, 1942, p. 6), all of which helped him to continue his studies at Owens after the death of his father. It was also during his time at Owens that he produced his first scientific paper, an experimental work to measure the electrical displacement when two non-conductors were put into contact, which was published in the *Proceedings of the Royal Society*, under the patronage of Balfour Stewart. This paper shows only the tip of the iceberg of the uniqueness of J. J.'s training before moving to Cambridge, for it demonstrates that Thomson had a much deeper scientific training, both in mathematical and in experimental physics, than most of his peers at the university. His exposure to experimental science would prove particularly significant after his years as an undergraduate in Cambridge, where all his training had an exclusively theoretical character, when, after graduating, he decided to complement his training with the acquisition of more experimental skills and he, eventually, eventually as Cavendish professor.

This particular early training would, however, prove to be a hindrance for entering Cambridge. Thomson failed at his first attempt due to his excessively scientific education, which had left some more basic areas, especially in the humanities, unattended. The feedback he got from the examination stated that he 'should have done better if, instead of reading the higher subjects in mathematics which were not included in the examination, [he] had concentrated on getting a "thorough grounding" in the lower ones'. The following is Thomson's later ironic description of what this 'thorough grounding' meant: 'reading the subjects included in the entrance scholarship examinations over and over again, and doing a great number of trivial examples in them'. Furthermore, 'in some cases the boys, who were to be sent in to compete for entrance scholarships, did little in the two years before the examination but to write out answers to papers set in previous examinations. Under this system the boys get more and more fed-up with mathematics the longer they are at it' (Thomson, 1936, p. 31). As Andrew Warwick showed, the system of training young boys by repetition of case studies and previous examinations used in most public schools had its roots in the Cambridge pedagogical system itself (Warwick, 2003a, pp. 254–64). Once again, let us emphasize that Thomson did not arrive in Cambridge from this public school tradition, but from the unique early training at Owens College.

1.5   The Unseen Universe

Before we move on to examine Thomson's time in Cambridge, let us look at another factor that was present during Thomson's many hours in the laboratory with Stewart. The mixture of smoke and humidity that permeated the atmosphere in industrial Manchester is a compelling image of one important aspect in the world-view of Victorian scientists: the world of matter was equally permeated by an entity – the ether – which was a major seat of energy and interactions, and the medium for the transmission of light. The ether was supposed to be weightless but, at the same time, rigid enough to transmit light waves. The question about the relationship between ordinary matter and ether, between matter and energy, was at its speculative peak in the second half of the nineteenth century, giving the ether some of the attributes of science fiction among the educated public. Science made its existence necessary; its characteristics made it open to mystery and to all manner of speculation.

Balfour Stewart used this cosmological idea to write, in 1875, together with Peter Guthrie Tait, a bestseller on natural theology, called *The Unseen Universe*. Thomson would recall that 'Stewart had a strong turn for metaphysics', which explained the publication of a book that 'was an attempt to find a physical basis for immortality' (Thomson, 1936, p. 22). Taking ether as the ultimate reality in Nature, Tait and Stewart tried to prove the immortality of the soul and the possible existence of many spiritual entities (but not the existence of a Creator, which they took for granted). The main idea was that the world as we know it, the 'visible universe' as they put it, was only a minor part, contingent and finite in time, of a greater universe, the Unseen Universe, which included all created things. 'We maintain that the visible universe – that is to say the universe of atoms – must have had its origin in time, and that while THE UNIVERSE is, in its widest sense, both eternal and infinite, the universe of atoms certainly cannot have existed from all eternity' (Stewart & Tait, 1875, p. 9). In this context, the atoms of matter would be a transient entity: 'We are not led to assert the eternity of stuff or matter, for that would denote an unauthorized application to the invisible universe of the experimental law of the conservation of matter which belongs entirely to the present system of things' (p. vii); or, to put it more bluntly, 'it appears no less false to pronounce eternal that aggregation we call the atom, than it would be to pronounce eternal that aggregation we call the sun' (p. vi).

Matter was regarded as a non-fundamental entity in the complete universe, but only as an ephemeral phenomenon of the visible universe. Here they introduce a distinction between 'objective' and 'substantive' reality, saying that, while atoms have both types of reality, the unseen world of ether is 'objective' but not 'substantive', an idea that can only be understood in the light of the science of energy that crystallized in the previous decades: 'It is only within the last thirty or forty years that there has gradually dawned upon the minds of scientific men the conviction that there is something besides matter or stuff in the physical universe' (Stewart & Tait, 1875, p. 100). And, continuing with the same kind of rhetoric, they take energy as this 'something' besides matter: 'Taking as our "system of bodies" the whole physical universe, we now see that … energy has as much claim to be regarded as an objective reality as matter itself' (p. 114–5). In Tait and Stewart's views, there was, however, an ontological asymmetry between matter and ether, for the latter was considered to be more fundamental than the former. This is consistent with such cosmologies as the one implicit in the vortex atom theory of William Thomson, which assumed atoms to be explainable in terms of vortices in the ether, and which became, in Kragh's expression, a 'theory of everything' in Victorian science (Kragh, 2002). However, they felt that they had to introduce a fundamental change in the conditions of the ether. In W. Thomson's theory, the primordial fluid – the ether – was seen as perfect, and the appearance of the vortex rings as the result of some external (divine) action on it. For Stewart and Tait, this would not accomplish the conditions of a self-sustained 'Unseen Universe'. Therefore, they regarded the ether as a non-perfect fundamental fluid, in which the vortices appear and disappear as a result of spontaneous fluctuations. In this way, the visible world would be ephemeral 'just as the smoke-ring which we develop from air … is ephemeral, the only difference being in duration, these lasting only a few seconds, and the others it may be for billions of years' (Stewart & Tait, 1875, p. 157).

This holistic idea was not characteristic only of Stewart and Tait: late nineteenth-century science was overenthusiastic about the possibilities of reducing all knowledge to one metaphysical principle from which all phenomena, including the spiritual, would be deduced (Harman, 1982; Myers, 1989; Smith, 1998; Noakes, 2005). *The Unseen Universe* is only one example of a 'growing commitment to a belief in the uniformity of nature, the restriction of divine action to the creation of the universe, the rejection of suppositions of divine interventions to explain apparent discontinuities in the natural world, and the separation of the natural and the supernatural' (Heimann, 1972, p. 75). The interest of this book, as far as J. J. Thomson is concerned, resides not only in the fact that it was a best-seller among those with interests in science and natural philosophy, but mainly in the fact that Stewart was writing this book precisely in the years when J. J. spent long hours in the laboratory under his guidance, and this must have certainly exerted a direct influence on him (Sinclair, 1987, p. 90). As Davis and Falconer stated, Thomson received from Stewart a thorough grounding in the prevalent Victorian method of reasoning by analogy and in ether physics (Davis & Falconer, 1997, p. 6; see also Chayut, 1991).

Moreover, as we shall see in Chapter 4, Thomson was directly involved in the Society for Psychical Research, a society devoted to the scientific study of paranormal phenomena, something to which some aspects of the physics of the ether were particularly suited. For the time being, it suffices to quote from a public lecture he gave in his hometown in 1907 in which he explained the relationship between ether and matter, between electricity and mechanics, in a language suited to his audience, mixing the physical, the mercantile and the mysterious:

> The study of the problems brought before us by recent investigations leads us to the conclusion that ordinary material systems must be connected with invisible systems which possess mass whenever the material systems contain electrical charges. If we regard all matter

as satisfying this condition we are led to the conclusion that the
invisible universe – the ether – is to a large extent the workshop
of the material universe, and that the phenomena of nature as we
see them are fabrics woven in the looms of this unseen universe
(Thomson 1907f, p. 21).

1.6   Undergraduate in Cambridge

While Owens College was a new institution born in a lively city
and promoted by the same entrepreneurial bourgeoisie that was developing Manchester's industry, Cambridge was a completely different world. The
model of science in that ancient and aristocratic university was mathematical
physics, the teaching of which was organized around the *Mathematical Tripos*.
Eminent Victorian scientists such as John Herschel, William Whewell, George
G. Stokes, William Thomson, Peter G. Tait and James Clerk Maxwell had all
been Mathematical Tripos students. Without completely denying the importance of observation and experimentation, the ideal science in Cambridge was
one in which data and theories had achieved a complete mathematical formulation, which could then be turned into the starting-point for the solution
to new problems. Whewell, the influential Master of Trinity College, worked
to emphasize the role of mathematical training in Cambridge in the mid-
nineteenth century. As he put it in 1837, the progress of the sciences 'depends
on the distinctness of certain fundamental ideas; and these ideas, being first
clearly brought into view by the genius of great discoverers, become afterwards the inheritance of all who thoroughly acquire the knowledge which is
thus made accessible' (Whewell, 1837, p. 20). The role of university training in
the sciences was, from this perspective, the transmission of fixed principles or
fundamental ideas, and the usual work was to deductively develop new consequences of such principles by means of reasoning and mathematical work
(Williams, 1990). With this model in mind, the experimental sciences appeared
as a kind of second-class knowledge. They were provisional and particular and
they lacked the rigour of mathematical formulation.

With the increasing specialization of the different sciences, however,
Cambridge accepted the need to create a new *tripos* of experimental science,
and, in 1851, the *Natural Science Tripos* was established. The chairs involved in this
tripos were, at first, chemistry, experimental natural philosophy, mineralogy,
geology, comparative anatomy, physiology, and botany. Physics, being at the
core of the Mathematical Tripos, was not, at the beginning, part of the Natural
Science Tripos, thus embodying the Whewellian distinction between adult and
under-aged sciences. Only with the establishment of the Cavendish Laboratory

did experimental physics enter the Natural Science Tripos, and, when Thomson went to Cambridge in 1876, physics was being taught, although from radically different perspectives, in both the Mathematical and Natural Science Triposes. However, it was still the case that the Mathematical Tripos had a much greater prestige in Cambridge, and it was the one that promising students like J. J. were expected to follow (Roberts, 1989).

Thomson arrived in Cambridge in October 1876 and, as he put it in his auto-biography, "kept" every term since then, and [was] in residence for some part of each Long Vacation' (Thomson, 1936, p. 34), which means that he never left Cambridge for more than a few weeks after his arrival at the age of 19. After his first failed attempt to enter Cambridge, Thomson prepared himself again for the examination and, on 18 April, 1876, he resat the examination to get a fellowship at the most prestigious of Cambridge institutions: Trinity College. Only the wealthiest colleges in Cambridge could afford to offer some scholarships for brilliant students with meagre financial resources like Thomson. The examination consisted of two papers 'confined to questions in Arithmetic, Geometry, Algebra, Trigonometry, Conic sections treated both geometrically and analyt-ically, the Elements of the Differential Calculus, and Mechanics as far as the Dynamics of a particle included' (CUR, 19 October 1875, p. 45), some of which involved a serious mastery of advanced mathematics. He was one of six candidates to win a 'Minor Scholarship', with a value of £75 a year. Just to give an idea of the number of students awarded a scholarship, in the year 1876, there were only 12 scholars out of 167 matriculating students at Trinity College (CUR, 21 November 1876). The following year he sat for a 'Foundation Scholarship', a fellowship open to undergraduates of the college, which increased both his income to £100 a year and his prestige in the college.

Life as an undergraduate in Cambridge was, and in some respects still is, a unique experience. Even though the introduction of the new *triposes*, with tighter syllabuses, was changing the face of the university, the long tradition of liberal education made itself felt very prominently. Cambridge was not only a place of academic learning, it was a factory of mass production of gentlemen to serve the British Empire. That is why the strictly academic education went hand in hand with a very intense social life, which included college celebrations, sports tournaments, religious events, and a wide range of activities in which the young men developed a particular *savoir faire* that turned them into exemplary Victorians. Thomson found this new world attractive and he did not seem to have many problems fitting in. At the time, the college fellowships provided their recipients with enough money to spend entertaining their friends and colleagues in college, something that, in time, he would consider to be essential for any Cambridge student.

Thomson described his days as an undergraduate as 'very pleasant but uneventful' (Thomson, 1936, p. 52), and, with a closer look at his recollections, we can infer that this means he was very much work oriented and not specially good at those activities undergraduates tended to value most: sports. The main games in those days were cricket and football, and rowing was just becoming increasingly popular. However, Thomson, like 'the ordinary reading man who was not particularly good at games, had not much chance of playing either cricket or football … there was no room for the "rabbits"' (p. 66). J. J., like most of his peers got the necessary physical exercise by taking walks: 'Between 2 and 4 in the afternoon, streams of undergraduates, two and two, might be seen on all the roads within three or four miles of Cambridge … On Sundays many went further afield and walked for five or six hours'. Occasionally he would play a tennis match or some golf, sports that 'only require two players and for which it is generally possible to find an opponent nearly as bad as yourself so that you don't feel you are spoiling anyone's game' (p. 67). The confidence and brilliance shown in his academic career certainly failed him in more prosaic activities.

Chapel attendance was compulsory for certain festivities and events. It is not easy to assess J. J.'s attitude towards religion. He came from an average Anglican family, and there is no reason to assume either a rejection of religious practice, or an excessive religious fanaticism. As an example, in his autobiography, he amusingly told how difficult it was, at times, to attend the 7.30 a.m. service, since 'though we got up early it was not quite early enough, and the Chapel doors were shut before we could reach them' (Thomson, 1936, p. 53). As for entertainment in Cambridge, there was not much to do in the town. The only theatre was open solely during Vacation, and the amateur drama society of the university performed two plays per year. As for dancing, the only such events were in May Week, partly due to the extremely low numbers of young ladies around the university. These arrangements meant that a student could seriously concentrate in his work, rather than getting distracted by other activities.

1.7   Second wrangler in the Mathematical Tripos

The first port of call in the education of a Cambridge student was the college tutor. He was responsible for guiding the student during his undergraduate years, looking after his welfare in all the aspects of his life in the university. The long tradition of liberal education in Cambridge meant that there was no fixed syllabus, but the students were free to choose the courses and lectures they wanted to attend. The tutor was responsible for guiding the student in his choices among a sea of college, intercollegiate, and departmental lectures. Thomson's tutor, Mr J. M. Image, was a classicist, with little idea about the intricacies of mathematics, and thus Thomson could choose the lectures that most suited his interests and needs and avoid some lecturers, 'the dullness of some of [which] can hardly be imagined' (Thomson, 1936, p. 44).

In Victorian Cambridge, however, candidates for the higher ranks in the Mathematical Tripos underwent an intensive training under the guidance of an independent coach, who made sure that his students read all the subjects included in the examinations, something not easy considering the large number of subjects and the limited time to study them. Thomson, like many of his contemporaries, was coached by the charismatic Edward Routh. The system used by the coaches consisted of combining their lectures with very competitive weekly targets in solving examination problems from previous years. The training also included time constraints: 'one week we could take as much time as we pleased in solving the problems, the next we were expected to do them in three hours, the time allowed for such a paper in the tripos'. The students sent their papers for marking and a few days later 'a complete solution of the paper in Routh's handwriting was placed in the pupil's room, together with a list of the marks each pupil had obtained. This introduced a sporting element, and made us take more trouble over them than we should otherwise have done' (Thomson, 1936, p. 38). This system of coaching created an army of mathematicians who were convinced that any problem in physics could, in the end, be studied using the powerful tools of modern mathematical analysis. The problem-solving pedagogy meant that the students only thought about how to solve the problem, not about whether the problem was soluble or not.

Routh had been senior wrangler in 1854, beating James Clerk Maxwell in the rankings. The following year, he was elected to a fellowship and lectureship at Peterhouse (one of the colleges in Cambridge), and, from that position, he became the most successful of Cambridge coaches: of the 990 wranglers who graduated between 1862 and 1888, almost half were coached by Routh (including 26 senior wranglers), while between 1865 and 1888, 80 per cent of the top three wranglers were his pupils (Warwick, 2003a, p. 233). This made him the most influential mathematician in Cambridge, far beyond any other professor or college tutor. Thomson vividly recalled Routh's lifestyle:

> The regularity of Routh's life was almost incredible; his occupation during term time could be expressed as a mathematical function of the time which had only one solution. I believe one who had attended his lectures could have told what he had been lecturing upon at a particular hour, and on a particular day, over a period of twenty-five

years. The fact that year after year he gave the same lectures at the same time did not make him stale as it would most people. He might, as far as one could judge from his manner, have been delivering each lecture for the first time. His way of taking exercise was as regular as his lectures: every fine afternoon he started at the same time for a walk along the Trumpington Road; went the same distance out, turned and came back. His regularity was not, as might perhaps have been expected, accompanied by formal and stereotyped manners; these were very simple and kindly and we were all very fond of him (Thomson, 1936, p. 40–1).

Historian Andrew Warwick has analysed the pedagogical tradition in Cambridge and the particular role played by Maxwell's *Treatise of Electricity and Magnetism*. Published in 1873, just after his appointment as the first Cavendish Professor, the *Treatise* was intended as a compendium of theoretical and experimental approaches to electricity and magnetism, and developed Maxwell's new views on the field. The *Treatise* was, thus, thought of as a textbook to aid him in his professorial role. In the hands of other people, however, making sense of the *Treatise* proved a difficult task, since there was a lot of tacit knowledge, which most readers would not necessarily have. That is why Cambridge students in the late 1870s were in a privileged position to understand Maxwell's work even though, when doing so, they developed one particular understanding of it. 'The collective understanding of electromagnetic field theory that emerged in Cambridge circa 1880 was shaped by a combination of the problem-solving approach … and discussions at the intercollegiate lectures on the *Treatise* held at Trinity College by W. D. Niven' (Warwick, 2003a, p. 291). Outsiders, on the other hand, lacking the pedagogical tools of Cambridge undergraduates, would find it very difficult to make sense of the *Treatise*. An example of this is the speed with which Thomson was able to develop important consequences of Maxwell's electromagnetic theory: 'Where men like Heaviside and Fitzgerald had taken years of private study to begin offering their own original contributions, Thomson wrote two important papers on electromagnetic theory within months of graduating while working simultaneously on several other projects' (p. 343; see Thomson, 1880 and 1881a).

After Routh, the aforementioned William D. Niven was the second most influential person in Thomson's education in Cambridge. Niven had been third wrangler in the Mathematical Tripos in 1866 and, in 1874, was invited to return to Trinity College as a lecturer in mathematics. In Cambridge he became a close friend of Maxwell, to the extent that he considered himself as his intellectual heir, especially after Maxwell's untimely death in 1879. Niven was responsible for the second edition of the *Treatise* as well as for editing the two volumes of Maxwell's scientific papers (Niven, 1890). Thus, he was the person to meet if one wanted to understand the *Treatise*. His intercollegiate lectures, attended by ambitious undergraduates and new wranglers, became the forum of discussion of the new ideas (Warwick, 2003a, p. 317). Thomson would remember that the importance of Niven's lectures resided not so much in their clarity but in his enthusiasm:

> Niven was not a fluent lecturer nor was his meaning always clear, but he was profoundly convinced of the importance of Maxwell's views and enthusiastic about them; he managed to impart his enthusiasm to the class, and if we could not quite understand what he said about certain points, we were sure that these were important and that we must in some way or other get to understand them. This set us thinking about them and reading and re-reading Maxwell's book, which itself was not always clear. This was an excellent education and we got a much better grip of the subject, and greater interest in it, than we should have got if the question had seemed so clear to us in the lecture that we need not think further about it (Thomson, 1936, p. 42–3).

William Niven became a lifelong friend of J. J. He 'was one of the best and kindest friends I ever had; he was very kind to me from the time I came up as a freshman. He often asked me to go for walks with him. I went very often to his rooms and, through him, I got to know many of the Fellows of the College' (Thomson, 1936, p. 43). One sign of this close and long lasting friendship with Niven is that he eventually became J. J.'s first son's godfather.

Among the other lecturers in his undergraduate years, Thomson would recall with particular affection J. W. L. Glaisher, whose lectures were 'the most interesting I ever attended' (Thomson, 1936, p. 44), not only for his clarity and enthusiasm in pure mathematics, but also for the amusing anecdotes he used to tell during his lectures. Other lecturers were Professors A. Cayley, J. C. Adams and G. G. Stokes. The first was significant for the particular way he had of solving problems: 'he did not seem to trouble much about choosing the best method, but took the first that came to his mind. This led to analytical expressions which seemed hopelessly complicated and uncouth' but after a while, 'in a few lines [he] had changed the shapeless mass of symbols into beautifully symmetrical expressions, and the problem was solved' (p. 47). Thomson saw this as a good lesson not to be afraid of complicated mathematical expressions. By contrast, Adams' lectures were magnificently clear and ordered, since he 'carried the feelings of an artist into his mathematics, and a demonstration had to be elegant as well as sound before he was satisfied' (p. 47). His only problem was that he read his lectures, which made them less appealing. Finally, Stokes' lectures were Thomson's favourites: 'for clearness of exposition, beauty and aptness of the experiments, I have never heard their equal' (p. 48).

The world of the *Treatise* was a world of ether. In the book, 'an attempt [was] made to explain electromagnetic phenomena by means of mechanical action transmitted from one body to another by means of a medium occupying the space between them' (Maxwell, 1873, 2, §781). The existence of the ether was technically only a hypothesis, but one that could be supported by different and independent theories. Light and electromagnetism, previously independent phenomena, were united under the same explanatory framework. Light was, in Maxwell's theories, a manifestation of electromagnetic waves. This unification was a boost for the existence of an ether that had first been postulated to account for the transmission of light. In Maxwell's words, 'if the study of two different branches of science has independently suggested the idea of a medium … the evidence for the physical existence of the medium will be considerably strengthened' (§781). The fact that electrical and optical evidence independently supported the hypothesis of an ether produced in Maxwell and those around him a 'conviction of the reality of the medium similar to that which we obtain, in the case of other kinds of matter, from the combined evidence of the senses' (§781).

Electromagnetic phenomena could be explained by considering the dynamical properties of the all-pervading ether, i.e., assuming the ether to be a continuous medium in which kinetic and potential energy could be stored. The transmission of this energy through the ether, consistent with the principle of conservation of energy, would account for all the electromagnetic processes. Even though Maxwell had, for certain purposes, a particular model for the structure of the ether, imagining it in terms of hexagonal cog-wheels in motion, this structure was not important to account for the electromagnetic phenomena. The only properties he needed the ether to have were those of a continuous medium (Buchwald, 1988). The equations valid in hydrodynamics would, then, be valid in describing the ether, in the same way that one does not need to assume liquids are composed of atoms in order to discuss the hydrodynamic properties of fluids.

The last sentence of the two-volume opus makes it clear that his primary goal in the *Treatise* was to understand the mechanisms by which the ether acts: 'if we admit this medium as an hypothesis, I think it ought to occupy a prominent place in our investigations, and that we ought to endeavour to construct a mental representation of all the details of its action, and this has been my constant aim in this treatise' (Maxwell, 1873, §866). Thomson's education was, as any Mathematical Tripos student in the 1870s, deeply imbued by this meta-physical world-view.

Overall, the education in physics received in the Mathematical Tripos was fundamentally theoretical. In J. J.'s times, students for the Mathematical Tripos were not expected to attend any demonstrations in the new Cavendish Laboratory. In fact, there had been attempts to include questions in the examinations the answer of which was only possible if one had some experimental knowledge; but the repeated failure to engage students and coaches alike in this project meant that such questions were systematically left unanswered by all students. By the late 1870s, 'it had become clear that mathematics students could grasp the principles of, say, Wheatstone's Bridge and its application to practical electrical measurements without actually witnessing the instrument in action' (Warwick, 2003a, p. 315). That explains why, in his undergraduate years, Thomson never entered the Cavendish or even met Maxwell. Nevertheless, Thomson did not see this almost exclusively mathematical training as a handicap for his career as a physicist: 'I have found this of great value (*c'est le premier pas qui coûte*), and it is a much less formidable task for the physicist, who finds that his researches require a knowledge of the highest parts of some branch of pure mathematics, to get this if he has already broken the ice, than if he has to start *ab initio*' (Thomson, 1936, p. 39).

Finally, in January 1880, Thomson sat the Mathematical Tripos examinations. The tension on examinations days was extreme. Even such a relaxed and self-confident man as J. J. Thomson had those days deeply engraved in his memory, especially the insomnia he suffered during the last five days. 'Insomnia', he recalled, 'is even more unpleasant in Cambridge than in most other places, for since several clocks chime each quarter of an hour you know exactly how much sleep you have lost, and this makes you lose more' (Thomson, 1936, p. 63). In spite of this difficulty, Thomson performed very well and ended up second wrangler, Joseph Larmor topping the list. The total number of people awarded honours was 99.

2

J. J. Thomson's early work in Cambridge: a continuous and all-embracing physics

2.1   In Cambridge as a graduate

J. J. Thomson decided to stay in science, and in the years immediately after his graduation, he worked intensely to complement his education in order to secure a position in academia. An obvious choice for a high wrangler was to try to stay at Trinity College with a fellowship, which he won in the summer of 1880. Another option arose in 1881 in the form of a vacant professorship of applied mathematics at his very own Owens College, but his application was turned down in favour of Arthur Schuster, who was more experienced and already teaching there. The fascinating thing about the period 1880–1884 is not only the amount of work he did, both theoretical and experimental, but also the scope of it: we find him developing a dynamical theory of physics and chemistry for his Trinity College fellowship dissertation, and a dynamical theory of matter with which he won the prestigious Adams Prize of 1882; we find him in the Cavendish, first trying to acquire the experimental skills that were deemed unnecessary for Mathematical Tripos students, and later undertaking some of the basic experiments that Maxwell had devised in the *Treatise*; and we also find him developing some basic theoretical work on electrodynamics. All this work finally gained him a university lectureship in Cambridge in 1883 and Fellowship of the Royal Society in the spring of 1884, even before he was appointed Professor of Experimental Philosophy and director of the Cavendish Laboratory.

Competition for college fellowships in Cambridge was fierce, and candidates were allowed to apply up to three times, in the first, second and third summers after graduation. It was rather unusual to make use of the first attempt since, besides sitting for an examination, candidates had to write an original piece of research work and possibly compete with people in their second or third year after graduation; but Thomson applied for it and, contrary to the expectations of his college tutor, he was elected to a fellowship. In retrospect, J. J. explained that the reason for his success was, partly, that his research topic had been in his mind, only waiting to be formally developed, since his days at Owens College with Balfour Stewart: the reduction of all forms of energy to manifestations of kinetic energy. Parts of this dissertation were later published in two papers in the *Philosophical Transactions* (Thomson 1885a, 1887), and formed the core of his first book, *Applications of Dynamics to Physics and Chemistry* (Thomson, 1888). This work linked his early ideas from Owens with the mathematical methods that he had learnt in Cambridge. In the same way that Maxwell had merged mechanical and electrical energies, Thomson thought it his task to extend this methodology to its logical conclusion: to reduce all forms of energy (potential, electromagnetic, chemical, thermal, and so forth) to kinetic energy (Topper, 1971).

This first project shows a characteristic that I want to emphasize in this chapter: Thomson had an all-embracing view of what he called the *Physical Sciences*, a concept that included all branches of physics and some areas of chemistry. In the preface to *Applications of Dynamics to Physics and Chemistry*, Thomson argued that, since the issues in the book 'relate to phenomena which belong to the borderland between two departments of Physics [meaning physics and chemistry], and which are generally either entirely neglected or but briefly noticed in treatises upon either, I have thought that it might be of service to students of Physics to publish them in a more complete form' (Thomson, 1888, preface). The first few chapters of this book introduced the reader to the mathematical methods of Lagrange and Hamilton as presented both by Maxwell and by Routh, which enable the mathematical physicist to deal with physical phenomena without knowing the intrinsic properties of the system or the nature of the mechanism to be investigated. The phenomena to which these methods were applied in the following chapters of the book included temperature, electromotive forces, elasticity, evaporation, absorption of gases by liquids, surface tension of solutions, chemical dissociation, chemical equilibrium and the connection between chemical change and electromotive forces.

Thomson heavily relied on the power of the mathematical methods that he had learnt in the Mathematical Tripos, which he thought were also valid to explain many chemical phenomena. To give just a few examples, when talking about chemical dissociation, Thomson claimed that 'this phenomenon has some analogy with that of evaporation', since one could imagine that in the same way there is an equilibrium between liquid and gas that allows for condensation and evaporation between the two states of matter, 'so in dissociation, we have also equilibrium between portions of the same substance in two different conditions', even though, in this case, both would be in the gaseous state: 'the molecules in the one condition being more complex than those in the other, and matter being able to pass from one condition into the other by the more complex molecules splitting up, "dissociating" as it is called into the simpler ones, while on the other hand some of the simpler ones combine and form the more complex molecules' (Thomson, 1888, p. 193). In a later chapter on electrolysis, the continuity between physics and chemistry is again made clear: 'the principle that when a system is in equilibrium the Hamiltonian function is stationary can be applied to determine the connexion between the electromotive force of a battery and the nature of the chemical combination which takes place when an electric current flows through it' (p. 265). Thus, Thomson's work tried, from the very beginning, to extend the dominions of his physics into the realms of chemistry. He not only thought that chemical processes were partly explained by physico-dynamical processes, but also that the tools to study chemical processes were the mathematical methods he had learnt in the Mathematical Tripos.

It has been said that this 1888 book 'made [Thomson] one of the founders of the new science of theoretical physical chemistry. If he had continued in it, he might have become as distinguished in physical chemistry as he became in physics' (Crowther, 1974, p. 107). Thomson's all-embracing programme was, however, rather different to the early discipline of physical chemistry. According to most historians of science, physical chemistry was the outcome of chemists interested in the physical foundations of chemical phenomena. The fathers of physical chemistry, the Swede Svante Arrhenius, the German Wilhelm Ostwald, and the Dutch J. H. van't Hoff, were chemists – mainly organic chemists – willing to develop a new discipline (Servos, 1990; Nye, 1993). Thomson, on the other hand, was mainly a physicist trying to extend the methods of physics into the chemical dominions, in a more all-embracing way. That may explain why Thomson was not in contact with the new physical chemists until well into the twentieth century, after Thomson's corpuscle became the electron of the physical chemists (Nye, 2001).

2.2   Early experimental work at the Cavendish

Although not completely ignorant of experimental matters, thanks to his time at Owens College, during the two and a half years of the Mathematics Tripos he visited the Cavendish only occasionally to see his Mancunian friends John H. Poynting and Schuster. Had he then spent more time there, he would have met Maxwell, the first director of the Cavendish, who died prematurely in 1879. The laboratory was under the direction of Lord Rayleigh at the time of Thomson's first reencounter with an experimental laboratory in 1881, a reencounter that was not totally promising. After a first unsuccessful attempt 'to detect the existence of some effects which I thought would follow from Maxwell's theory that changes in electric forces in a dielectric produced magnetic forces' (Thomson, 1936, p. 97), Rayleigh suggested that Thomson should start with easier, more standard, experiments that would allow him to acquire a close familiarity with basic experimental techniques in electromagnetism. In a few months, however, he had gained enough experience to publish his results in his first Cavendish paper in the *Philosophical Magazine* (Thomson, 1881b).

These experiments confirmed that 'a coil acts in many cases like a condenser, and possesses appreciable electrostatic capacity', something that had been incidentally mentioned by Helmholtz and others, and about which 'I am not aware, however, of any experiments in which this property produced any very marked effect' (Thomson, 1881b, p. 49). Thomson's aim was, basically, not so much to prove this effect, but to measure its intensity. From a practical point of view, the experimental set-up was nothing new, but closely relied on some electromagnetic experiments that Lord Rayleigh had done ten years earlier: two wires were wound side by side on a bobbin, one being the primary and the other the secondary circuit on which the current was induced. The goal of the experiment was to measure the condenser capacity of the secondary coil by trying to find induced currents in it when the current in the primary coil was broken. In order to do so, this second coil was cut into sections that were linked by small spirals through which the expected induced current passed. A small magnetized needle inside this spirals measured the intensity of such induced currents and, indirectly, the condenser capacity of the coil. This work was relatively simple and straightforward, but helped Thomson familiarize himself with the Cavendish regime of experimental work in electromagnetism that had been initiated by Maxwell and which Lord Rayleigh was continuing. I shall comment on this tradition later in this chapter.

His next experimental work, also suggested by Lord Rayleigh, was published two years later, in 1883, time by which he had been able to move towards more complicated experiments. One of the most dramatic consequences of Maxwell's work was the suggestion that light was an electromagnetic disturbance propagated in the same medium through which other electromagnetic actions are transmitted. This relationship between electromagnetism and light would eventually be explicitly observed in the 1888 experiments of Heinrich Hertz. At the time of writing the *Treatise*, Maxwell thought that this relationship could be inferred from the value for the speed of light in air V, and also by measuring 'v', the ratio between the electrostatic and electromagnetic units of electric charge, which had dimensions of a speed, and which physically accounted for the speed of the transmission of electromagnetic waves. Since the ways to obtain both values were totally independent, 'the agreement or disagreement of the values of V and of v furnishes a test of the electromagnetic theory of light' (Maxwell, 1873, §786). The *Treatise* also gave a number of possible experiments that would provide a value of v.

By 1880 there had been up to six attempts to measure v, all of which gave results of the same order of magnitude as the speed of light. This latter value was also only known in first approximation. Neither value could 'be said to be determined as yet with such a degree of accuracy as to enable us to assert that the one is greater or less than the other. It is to be hoped that, by further experiments, the relation between the magnitudes of the two quantities may be more accurately determined' (Maxwell, 1873, §787). And this was the fundamental task to which Thomson was about to contribute in his first serious experimental paper (Thomson, 1883b). His work followed one of the methods for determining v suggested by Maxwell: a modified Wheatstone bridge with a condenser in one of its arms was used to measure simultaneously the capacitance of the condenser (an electrostatic unit) and the resistance of the different arms of the bridge (an electrodynamics unit). In his biography of Thomson, the son of Lord Rayleigh pointed at two facts about these experiments worth mentioning. First, the then director of the Cavendish 'had already designed some of the apparatus to be used, and had contemplated taking part in the work himself … but … Thomson rather ran away with it, a natural result of energy, enthusiasm and self-reliance'. Second, Thomson seems to have been 'over sanguine' in spite of his 'rather limited experience in experiments', overlooking possible sources of error 'without applying the test of using alternative methods' (Rayleigh, 1942, p. 18). The result he obtained, 2.963×10^{10} cm per second, was the closest to the speed of light known at that time (around 2.998×10^{10} cm per second), but it was later proved to have an accumulated error of 1%. In 1890, he repeated the experiments with G. F. C. Searle, identifying some of the sources of error, and obtained a value of 2.995×10^{10} (Thomson & Searle, 1890). As we shall see, by that time, J. J. had almost finished preparing the revised third edition to Maxwell's *Treatise*.

2.3   The origins of the electromagnetic theory of matter

Warwick described Thomson as one of the most prominent Maxwellians of the 'second generation', by which he meant the group of Cambridge graduates, including Thomson, Poynting, R. Glazebrook, and J. Larmor, who received a systematic introduction to Maxwell's work with

Routh as coach and W. D. Niven as lecturer. Thomson mastered the content, the ethos and the limitations of the *Treatise* to such an extent that he could 'fashion himself as the University's leading expert on Maxwell's theory' (Warwick, 2003a, p. 334). The training in the Mathematical Tripos was such that only a few months after graduating, and while preparing his dissertation for the Trinity College fellowship, J. J. managed to publish a theoretical paper in the *Philosophical Magazine* in which he went beyond the *Treatise* in describing the behaviour of light as an electromagnetic wave.

The object of this paper was to generalize Maxwell's equations of light 'by taking into account the motion of the medium through which light is passing' (Thomson, 1880, p. 284), which could help validate Maxwell's theory of light with specific experimental results in different media. In particular, Thomson's generalized equations of light helped him to derive mathematical formulae for the reflection and refraction of a ray of light at the surface of a transparent medium, as well as the change in speed for light in a moving medium. As Warwick pointed out, 'where Maxwell had attempted to develop a complicated dynamical account of the interaction of electromagnetic waves with transparent media, Thomson assumed only that the rapidly oscillating electromagnetic fields of a light ray were subject to the same boundary conditions at the surface of a dielectric as were static electric and magnetic fields' (Warwick, 2003a, p. 337). The results Thomson obtained were in accordance with the ones given by the optical theory of light. J. J. was here embodying the ethos of a Maxwellian of the second generation: on the one hand, he was focusing on the central question of the *Treatise*, i.e., the electromagnetic nature of light, as he had also done in his first serious experimental work; on the other hand, he was addressing the problem with the use of Lagrangians, and not with the use of some intricate mechanism for the behaviour of the light ray, as Maxwell himself had unsuccessfully tried in the 1860s. Furthermore, by applying the continuity of boundary conditions to the case of rapidly oscillating fields, Thomson was basically putting together two different sections from the *Treatise*. This was the 'kind of information that a member of the first generation [of Maxwellians] might have passed on to a keen student like Thomson'.

More relevant for the history of electrodynamics was his paper 'On the electric and magnetic effects produced by the motion of electrified bodies', published in April 1881, and normally mentioned as the foundation stone of the so-called electromagnetic theory of matter (Thomson, 1881a). Again, this paper aimed at developing Maxwell's *Treatise* further, but now the trigger was some recent experiments on cathode rays by William Crookes and by Eugen Goldstein, who at that time thought that these rays consisted of charged particles at high speed (Darrigol, 2000, pp. 274–87). J. J.'s paper consisted of a mathematical analysis of the behaviour of a charged particle in an electric and magnetic field. In particular, he studied three aspects: 'the force existing between two moving electrified bodies, what is the magnetic force produced by such a moving body, and in what way the body is affected by a magnet' (Thomson, 1881a, p. 229). The particular way Thomson dealt with the mathematical formalism in this paper is, yet again, a very good example of his training in the Mathematical Tripos in the late 1870s: far from making new and complex mechanical suppositions, Maxwell's equations were treated as a given that had to be fulfilled.

The starting point for J. J. was the case of a charged sphere moving in a medium of a certain inductive capacity. Following Maxwell, Thomson argued that the moving charged sphere would induce a variation of the electric displacement at every point of the medium. 'Now, according to Maxwell's theory', he wrote, 'a variation in the electric displacement produces the same effect as an electric current; and a field in which electric currents exist is a seat of energy; hence the motion of the charged sphere has developed energy, and consequently the charged sphere must experience a resistance as it moves through the dielectric' (Thomson, 1881a, p. 230). This resistance could be mathematically treated as an increase of mass of the moving particle. Thus, the mass of the charged particle increases with its own movement, like any solid moving in a perfect fluid. Following the analogy with a solid sphere moving in water, 'when the sphere moves it sets the water around it in motion … This makes the sphere behave as if its mass were increased by a mass equal to half the mass of a sphere of water of the same volume as the sphere itself. This additional mass is not in the sphere but in the space around it' (Thomson, 1936, p. 93).

The idea that a moving charged particle generated an increase of mass due to its own movement was picked up a decade later by his closest rival in the Mathematical Tripos, Joseph Larmor, as well as others, both in England and on the Continent, to develop a theory of matter in which the origin of *all* mass was supposed to be of electromagnetic origin (Jammer; 1961, McCormmach; 1970, Darrigol, 2000). As we shall see in Chapter 4, in the first years of the twentieth century, J. J. Thomson himself would toy with the idea of electromagnetic mass as the source for the mass of his corpuscle and, since he was hoping for all matter to be composed only of corpuscles, for all mass whatsoever. The idea was then short-lived; but in 1881, and in spite of this paper, Thomson had no serious views on the possibility of a totally electromagnetic mass. As Jammer remarked many years ago (Jammer, 1961, p. 137), the example Thomson picked to illustrate the increase of mass in his 1881 paper shows how far he was from seriously considering any sort of an electromagnetic theory of mass. 'To form some idea of what the increase of mass could amount to in the most favourable case', he said, 'let us suppose the earth electrified to the highest potential possible without discharge, and calculate the consequent increase in mass' (Thomson, 1881a, p. 234). The calculations for this 'most favourable case' gave an increase of 650 tons for the earth, which is negligible compared to the whole mass of the earth. This case was certainly the 'most favourable' because this extra mass depended directly on the radius of the sphere, so the bigger the sphere the greater the increase in mass. With this, Thomson implied that the induced mass, although mathematically interesting, was actually irrelevant for ordinary matter.

To finish with this paper, let us go back to Thomson's motivation for it, which was to shed some light on discussions about the nature of the phenomena of discharge in vacuum tubes that Crookes and others had been studying. In particular, Crookes had started a controversy with German scientists about the nature of cathode rays, and whether this discharge phenomenon was better explained by corpuscular or undulatory theories. Furthermore, Arthur Schuster had been working at the Cavendish on discharge-tubes related phenomena, on which he would become an expert in Owens College. As we shall see, cathode rays would eventually become associated with Thomson's scientific career.In 1881, however, they were only one of many areas in which the ideas from the *Treatise* could be tried. Sparking from his study on the behaviour of electrified moving bodies, Thomson suggested a mechanism to explain one aspect of the discharge of cathode rays, i.e., 'the green phosphorescence observed in vacuum-tubes at places where the molecular streams strike the glass' (Thomson, 1881a, p. 237). He assumed that the collision of an electrified molecule on the glass of the tube would involve a reversal of the velocity of this charged body, therefore a very large variation in the vector potential generated by it and, as a consequence, the glass would be subjected to a rapidly varying electromagnetic force. But this, 'if Maxwell's electromagnetic theory of light be true, is exactly what it is subjected to when a beam of light falls upon it, which we know is the ordinary method of exciting phosphorescence'. This last statement again illustrates something we have already found: Thomson's early work after graduation was typical of a Cambridge Maxwellian of the second generation in that, while taking the *Treatise* as the basic electromagnetic theory to be developed, he was also aware of its basic limitation, i.e., the need for a clear confirmatory experimental proof of the relationship between light and electromagnetic fields.

2.4   The vortex ring theory of the atom

Perhaps the most revealing work of Thomson in this early period is his essay 'On vortex rings' (Thomson, 1883a), with which he won the prestigious

Adams Prize in 1882, a prize that had been established in 1848 to commemorate the discovery of Neptune by John Couch Adams, and which was, at the time, open to Cambridge graduates only. The subject for that year was 'an investigation of the action of two vortex rings on each other', a topic that was typical of the Cambridge of the day and fashionable among many British mathematicians and physicists (Kragh, 2002). The question of vortex rings had been a topic of interest since 1867, when William Thomson had suggested an atomic model in which atoms could be thought of as vortex rings in the ether. Besides its unifying character, this theory of a plenum was interesting since it maintained the immutability of primordial atoms. Helmholtz had shown that vortex filaments in a perfect fluid would not dissipate or be destroyed. Since the ether was mainly understood as a fluid, these results served to account for the indestructibility of atoms, and, at the same time, to treat them as special manifestations of the ether. This conception had considerable impact among scientists, and in the 1870s and 1880s, 'British physicists became increasingly attracted to this simple picture of atomic matter involving a concentration of ether spinning like a smoke ring in air' (Topper, 1980, 41; see also Klein, 1973).

William Thomson, as the father of the vortex atom model, felt that the theory was consistent with two of his key philosophical prejudices: his enthusiasm for dynamical models and his profound dislike for atomism, understood as Lucretius or Newton did, i.e. 'the monstrous assumption of infinitely strong and infinitely rigid pieces of matter' (W. Thomson, 1867, p. 15). This model gave him, for some time, the possibility of explaining those atoms in terms of a more fundamental continuous fluid. W. Thomson disliked the atomic theory because he felt that 'Lucretius' atom does not explain any of the properties of matter without attributing them to the atom itself. ... Every ... property of matter has ... required an assumption of specific forces pertaining to the atom, (pp. 15–16). In other words, W. Thomson felt that the nature of matter was not fully explained by atomic theories: atoms could explain the organization and some of the properties of bodies, but neglected the explanation of what matter actually is. In the years up to 1880, W. Thomson worked at this hypothesis and tried to explain with it many physical phenomena, including gravitation, the kinetic theory of gases, the dissipation of energy, and the wave motion in solids and liquids (Smith & Wise, 1989). Although, by 1882, W. Thomson and P. G. Tait had given up this cosmological idea, the topic remained of interest to mathematicians. This explains why most of the papers on vortex theory were published in journals of mathematics, not physics journals, since it was regarded as a most interesting mathematical problem; not only hydrodynamics was involved, but also the new area of topology of knots (Kragh, 2002, p. 46).

In this milieu, the 1882 Adams Prize essay was intended mainly as an exercise with a purely mathematical interest. But J. J. Thomson managed to broaden the question once again and to turn the problem of the stability of two vortex rings into an all-embracing theory of matter, thus reviving the theory of vortex atoms as a true theory of matter. In his words, an atomic theory based on the behaviour of vortex rings 'has *á priori* very strong recommendations in its favour'. And he used the following analogy between vortex rings and the atomic theory:

> For the vortex ring obviously possesses many of the qualities which a molecule that is to form the basis of a dynamical theory of gases must possess. It is indestructible and indivisible; the strength of the vortex ring and the volume of liquid composing it remain for ever unaltered; and if any vortex ring be knotted, or if two vortex rings be linked together in any way, they will retain for ever the same kind of be-knottedness or linking. These properties seem to furnish us with good materials for explaining the durable qualities of the molecule (Thomson, 1883a, §1).

Moreover, many other properties of the molecules, either independently or in relation to other molecules, could be accounted for using this model. The theory could explain atoms as secondary structures of a primary entity, the ether, and so it could be regarded as a more fundamental explanation of matter, since 'it proposes to explain by means of the laws of Hydrodynamics all the properties of bodies as consequences of the motion of this fluid. It is thus evidently of a very much more fundamental character than any theory hitherto stated' (Thomson, 1883a, §1). The argument of simplicity was a very powerful one; the only fundamental entity of nature would be the all-present fluid and everything could be represented by the energy at every point. Many years later, in his autobiography, Thomson looked back to this theory with some nostalgia to say that 'there was a spartan simplicity about it. The material of the universe was an incompressible perfect fluid and all the properties of matter were due to the motion of this fluid' (Thomson, 1936, p. 94).

Thomson's essay 'On vortex rings', besides being a typical example of Cambridge mathematical work with its over 50 pages of Lagrangians and differential equations, reveals in its last section that all these calculations 'would enable us to work out a complete dynamical theory of gases' (Thomson, 1883a, §51), by which he meant both the physical and chemical properties of material substances. In fact, this was not his first encounter with vortex rings. While he was at Owens he had shared laboratory hours with Stewart at the time of the latter's writing of *The Unseen Universe* (Stewart

& Tait, 1875). Also at Owens, Reynolds was physically experimenting with vortex rings, and these images must have impressed a young J. J. Thomson (Falconer, 1985, p. 17). In the essay, he also studied the possibility of permanent combinations of elementary vortex rings, and concluded that there could be stable combinations in systems consisting of up to six such rings. This was in agreement with the possible valencies of most elements, and he was led to speculate as follows:

> The atoms of the different chemical elements are made up of vortex rings all of the same strength, but some of these elements consist of only one of these rings, others of two of the rings linked together, others of three, and so on; thus, in this case, each vortex ring in the atom would correspond to a unit of affinity in the chemical theory of quantivalence (Thomson, 1883a, §54; see Sinclair, 1987, p. 95).

In this model, the mass of the atoms is no longer their fundamental characteristic, and their chemical affinity assumes such a role, bringing fundamental chemical properties into the picture, which shows us again, from a different angle, that J. J. was very much interested in chemical combinations of elements and substances as he saw this as an approach to the better understanding of the constitution of matter.

Thomson managed to bring the study of vortex rings further into the realm of ultimate explanations of physical phenomena, so much so that, some months after he wrote the essay, he applied some of its results in an attempt to explain the conduction of electricity in gases (Thomson, 1883c). Furthermore, the essay 'On vortex rings' is also revealing about the way in which Thomson approached chemistry: he was trying to explain chemical processes in terms of dynamical physics, thus abandoning the idea of affinities as some sort of force that was different from mechanical forces. The fact that Thomson was trying to incorporate some aspects of chemistry into physics was also evident to those who read the essay in Cambridge. For example, G. H. Darwin, the Professor of Astronomy, congratulated Thomson on winning the Adams Prize in the following terms: 'The problems you have solved are of amazing difficulty, and the results of the greatest interest. May you go on and discover a true dynamical theory of chemistry' (Darwin to Thomson, 25 January 1883, CUL, Add. 7654, D4). Nevertheless this aim was not exclusive to Thomson, and indeed, in 1885, G. F. FitzGerald corresponded with Thomson while trying to develop a model of the electromagnetic ether, saying: 'I thought it possible that electrical forces might be explained by these general effects of vortices &c. and that chemical forces might be due partly to these and partly to actions produced by the distortions of the vortices' (Figure 2.1). And he added: 'For though chemical and electrical

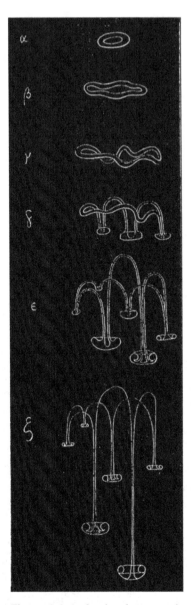

Figure 2.1 At the time he was seriously considering his vortex ring theory as an all-embracing theory of matter, Thomson also studied the properties of actual rings in a fluid (Thomson & Newall, 1885, p. 429). Courtesy: The Royal Society.

forces are due to like causes nevertheless chemical action is of a much higher order of complexity than simpler electrical actions (FitzGerald to Thomson, 1 January 1885, CUL, Add. 7654, F15). From his very early years as a researcher, Thomson proved that he had a deep interest in formulating a dynamical theory of chemistry, which would incorporate chemistry within the deductive physical sciences; and his work was seen in this way by many of his peers.

In his study of the vortex atom theory in Victorian science, historian Helge Kragh gave it a very definite lifetime: from 1867, the year of W. Thomson's suggestion, to 1898, when William Hicks, a professor in Sheffield, gave up the project. J. J. Thomson worked on it until 1891, when he moved to his theory of Faraday tubes. The new theory, which we shall encounter in the next chapter, was less cosmological, for its principal goal was to explain the interaction between electricity and matter, the key idea of his long research project studying the discharge in tubes. Eventually, that new theoretical framework led him to the discovery of the corpuscle. Faraday tubes, nonetheless kept one of the main features of the vortex atom theory, i.e., the assumption that the fundamental entity in nature is the ether and that atoms (and, later, corpuscles), are an epiphenomenon of the ether.

2.5   Director of the Cavendish Laboratory

In 1874, the Cavendish Laboratory was inaugurated at the New Museums Site, a piece of land that had previously hosted the Cambridge's Botanic Garden. Its main purpose was to serve as a teaching laboratory for the study of heat, electricity, and magnetism. These three subjects had been introduced into the Mathematical Tripos in 1868 and, as a result of this decision, the new laboratory of physics and a new chair of experimental physics were created. The first Professor of Experimental Physics was James Clerk Maxwell, who was capable of both highly sophisticated theoretical work and first-class experimental research. These two qualities were essential for the first Cavendish Professor in Cambridge, since they satisfied both traditionalist Cambridge dons, who thought that physics was necessarily mathematical, and also the reformers, who believed in a closer connection between the university and the practical, industrial and technological needs of the nation (Schaffer, 1992; Warwick, 2003a, pp. 264–72). The main role of the Cavendish was to serve as a laboratory devoted to teaching purposes. As was usually the case in Cambridge, in the early and mid-Victorian period, research was a professor's private activity. In fact, among the leading universities Cambridge was relatively late in incorporating a physics laboratory, so much so that the Cavendish was not part of the first generation of institutions in the so-called 'laboratory revolution' in Britain (Gooday, 1990; see also Larsen, 1962; G. P. Thomson, 1964; Crowther, 1974; Kim 2002).

In 1884, the chair of experimental physics in Cambridge was vacant. Lord Rayleigh had accepted the professorship after Maxwell's untimely death in

1879 but, after serving for over four years, he resigned, and the university had to start the search for a new professor. The candidates for the chair were Arthur Schuster (Manchester), Osborne Reynolds (Manchester), Richard T. Glazebrook (demonstrator at the Cavendish with Rayleigh), Joseph Larmor (first wrangler in 1880, now in Galway) and J. J. Thomson (Trinity College, Cambridge). The powerful Trinity College was eager to dominate the election, in spite of the fact that Rayleigh's favourite was Glazebrook, who had been at the Cavendish since 1876, and had thus worked with both Maxwell and Rayleigh. But precisely because of that, it is quite likely that some of the electors wanted a complete change at the Cavendish, for Maxwell and Rayleigh had connected it too closely to the electrical industry, a characteristic which was not appealing to the Cambridge establishment (Schaffer, 1992).

There has been much speculation about the reasons why the relatively inexperienced J. J. Thomson was appointed to the chair of experimental physics. Thomson said that his election as Lord Rayleigh's successor was a complete surprise to himself as well as to others (Thomson, 1936, p. 98). To take one example, when Schuster complained that a junior person had been appointed, his mentor in Manchester, Roscoe, tried to calm him down by pointing at the secrecy of any election in Cambridge: 'I do not at all agree with you that any slight is thrown upon the work of the Senior Candidates by the choice of a junior. The election is a very complicated function ... [and] the Electors swore a dreadful oath not to reveal anything whatever about his election' (Roscoe to Schuster, 25 December 1884, RSA, AS/B/163). However, it was not very unusual to appoint brilliant young students to professorships in Cambridge and the influence of Trinity College was strong enough to make the appointment of a Trinity man quite foreseeable. 'Well, no one can say that the appointment is a bad one. Only ... it might perhaps have been a better one', Roscoe said to the two Manchester contestants, Schuster and Reynolds, the day after the election (Roscoe to Schuster, 23, December 1884, RSA, AS/B/162). After his election, the vice-chancellor of the university, Norman MacLeod Ferrers, made the following statement: 'Professor J. J. Thomson combines a great amount of mathematical knowledge and power with, as I am assured, an experimental skill which promises to make him in the long tenure of office to which I trust he may look forward a worthy successor of the two distinguished men by whom the Cavendish Professorship has been occupied' (CUR, xv, 1885, p. 324). It is interesting that Ferrers regarded the experimental skills of J. J. Thomson as *promising*, as something that could be improved in the near future. It seems that the short time he had spent at the Cavendish between 1881 and 1884 was sufficient proof of his ability to become a good experimentalist, which, together with his mathematical skills, made him a promising replacement for Maxwell (Warwick, 2003a, p. 343).

A few of the first steps that Thomson took after his appointment are quite significant, for they also showed that he thought some aspects of chemistry relevant to the work of a laboratory of experimental physics. Glazebrook and William Napier Shaw, demonstrators at the Cavendish under Rayleigh, stayed in the laboratory. This was essential, to keep a certain continuity in the institution, especially in terms of teaching obligations, but they did not collaborate with Thomson in their research. A few months after his election, Thomson appointed Richard Threlfall, who had recently graduated in the Natural Science Tripos and had later spent some time in Rudolf Fittig's chemical laboratory in Strasbourg, as a demonstrator in physics. With him, he started a series of experiments on the chemical composition of gases. The first paper that they published was 'Some experiments on the production of ozone', which is clearly a chemical topic, and it was followed by a series of similar joint papers (Thomson & Threlfall, 1886). That these experiments can be classified as chemical is not an anachronism, for Thomson himself spoke of them in such terms: 'I am at present', he wrote to Schuster in early 1885, 'experimenting on a chemical thing, viz. the proportion of ozone formed at different pressures. I worked out the thing theoretically and am now trying whether the theory is right or not' (Thomson to Schuster, 1 March 1885, RSA, AS/C/331).

Threlfall's collaboration helps us to illustrate two aspects of Thomson's early days in his chair. On the one hand, he was clearly lacking in experimental skills and needed the help of people with more experience in the laboratory; on the other hand, he was determined to make the Cavendish a centre for *all* the physical sciences, chemistry included, thus abandoning the almost exclusively electromagnetic orientation which the Cavendish had had in the previous decade, which had turned the laboratory into what Simon Schaffer called a 'manufactory of Ohms' (Schaffer, 1992). In this respect, it is not wholly justifiable to assert that Thomson's lack of a research interest was the reason a wide range of apparently disconnected research projects were carried out at the Cavendish in the 1880s and early 1890s (Kim, 2002). If Thomson's idea of physics was all-embracing and included traditionally chemical topics, then all these researches had their place in the Cavendish. Threlfall was mainly an experimental chemist, and when he left for Australia in 1886, Thomson's experiments ran into trouble for lack of experimental know-how. He therefore often turned to people from the Chemistry Department for advice (Sinclair, 1987, p. 97). His letters to Threlfall speak about the advice he received from the Professor of Chemistry, George Liveing,[1] and some of his published papers

---

[1] Thomson to Threlfall, 7 Aug. 1886, CUL, Add. 7654, T19: 'Then, on Liveing's recommendation I tried acid but this nearly all disappeared when the tube was heated'.

acknowledge the help of the praelector in chemistry at Gonville and Caius College, Matthew M. Pattison Muir.

2.6   **Third edition of Maxwell's** *Treatise*

Thomson's own experimental work transcended the traditional boundaries between physics and chemistry. Shortly after his election in 1884, Thomson started a long-term project on the study of electrical discharges in tubes filled with gases. In the *Treatise*, Maxwell had said that 'these, and many other phenomena of electrical discharge, are exceedingly important, and when they are better understood they will probably throw great light on the nature of electricity as well as on the nature of gases and the pervading space. At present, however, they must be considered as outside the domain of the mathematical theory of electricity' (Maxwell, 1873, third edition (1891), p. 63). Thomson agreed that the study of discharge in gases promised to give insight into the nature of electricity (a physical problem) and the composition of matter (a chemical issue); but he thought that the time had now come to try to give a dynamical account of these phenomena. Dealing with discharge in tubes meant not only dealing with electricity, but also working with different gases, the preparation of which was clearly a task for chemists. Following the early theories of Crookes and Schuster, electrolysis was the model that he used to account for the phenomena that he observed in the discharge tubes, emphasising that molecules in the gas split to make the transfer of charge possible. The importance given to electrolysis can also be traced back to another suggestion of Maxwell in his *Treatise*, where he stated that 'of all electrical phenomena electrolysis appears the most likely to furnish us with a real insight into the true nature of the electric current, because we find currents of ordinary matter and currents of electricity forming essential parts of the same phenomenon' (Maxwell, 1873, third edition (1891), p. 58). Thus, Thomson's actual research project in the 1880s shows that he thought that the time had come for chemistry to be given the status of a grown-up science and thus to become a full member of the physical sciences. He believed that it was time for scientists to develop physical – dynamical – theories to account for the chemical problems of electrolysis, gas tubes, the constitution of matter, and the composition of gases.

Maxwell, the first Cavendish Professor, also had his own views on the boundaries between physics and chemistry. Whereas Whewell had made it clear that chemistry was not yet a proper physical science, Maxwell admitted that some areas of chemistry could be regarded as physics. In his classification, 'What is commonly called Physical Science occupies a position intermediate between the abstract sciences of arithmetic algebra and geometry and the morphological and biological sciences', where the morphological sciences 'are rich in facts, and will be well occupied for ages to come in the coordination of these facts' (Maxwell, in Harman, 1995, p. 777).

The case of chemistry was, for Maxwell, rather odd. Being a physical science, chemistry incorporated dynamical explanations but was expanding so fast that the theorists could not keep up with new developments. In his words, 'though [physical] Dynamical Science is continually reclaiming large tracts of good ground from the one side of Chemistry, Chemistry is extending with still greater rapidity on the other side, into regions where the dynamics of the present day must put her hand upon her mouth'. And he goes on saying that, however, 'Chemistry is a Physical Science, and that of very high rank. I do not, however, pretend to be able to go over its possessions and to show strangers the boundaries' (Maxwell, in Harman, 1995, p. 782).

Here, I emphasize the link between Thomson's and Maxwell's ideas because, at this point, J. J. was not only a prominent member of the second generation of Maxwellians but, in a way, he shaped himself as the natural continuator of the great physicist's, his successor in the chair of experimental physics, and the one who would bring Maxwell's project to its ultimate fulfilment. That partly explains his constant reference to the *Treatise* in most of the projects he undertook. This also gives us the context of one of his early tasks as director of the Cavendish: to prepare a third edition of the *Treatise*. A second edition of the complicated two-volume opus had been published in 1881. Maxwell himself had begun the task of correcting, changing and clarifying many aspects of the *Treatise*. At the time of his death, he had only had the chance to revise, correct and, at times, totally rewrite the first nine chapters. The rest remained unchanged except for some corrections of obvious mistakes by W. D. Niven, with the help of his brother Charles and J. J. Thomson. By the end of the 1880s, a new edition that incorporated clarifications and the latest developments in electromagnetism was needed, and J. J. agreed to take on this task. In the preface to this third edition of 1891, Thomson explained his original idea for this undertaking:

> When I began to revise this Edition it was my intention to give in foot-notes some account of the advances made since the publication of the first edition, not only because I thought it might be of service to the students of Electricity, but also because all recent investigations have tended to confirm in the most remarkable way the views advanced by Maxwell. I soon found, however, that the progress made in the science had been so great that it was impossible to carry out this intention without disfiguring the book by a disproportionate quantity of foot-notes (Thomson, in Maxwell, 1891, preface).

The solution he chose was to leave the *Treatise* as it was and to 'complement' it with a whole new volume that included the latest developments in electromagnetism and which is referred to throughout the third edition as the 'Supplementary Volume'. We will discuss this volume at the beginning of the next chapter. Now, let us go back to the institutional *tour de force* that Thomson undertook with the other scientific departments in Cambridge at the beginning of his tenure.

2.7   Mapping the domains of the physical sciences

From an institutional point of view, physics and chemistry were two quite separate worlds in Cambridge in the 1870s. Physics was mainly part of the Mathematical Tripos and chemistry was one of the central disciplines in the Natural Science Tripos. This state of affairs was not unique. According to John W. Servos (1990), by the middle of the century, chemists and physicists in many of the leading European universities worked in different institutes, used different instruments, and measured different properties. They even spoke different languages, since, whereas the chemist needed only arithmetic to express weight relations, the student of physics was becoming ever more dependent upon higher mathematics. Looking at the institutional development of the departments of physics, chemistry and, later, engineering, it could be said that Cambridge was just one particular case of this general description. However, looking at the work and the interests of some of the people who were influential in the university, one can argue that the process of specialization took place alongside a parallel – but unsuccessful – effort to unify the sciences, with mathematical physics as the exemplary science. As we have seen, this was the view of J. J. Thomson, but it was also that of the Professor of Chemistry, George Downing Liveing.

Liveing was the Professor of Chemistry in Cambridge from 1861. His education made him capable of teaching the natural sciences in the Cantabrigian style. He graduated as eleventh wrangler in 1850 and enrolled in the first ever Natural Science Tripos course, after which he was awarded a lectureship at St John's College. The college soon realized the need for new facilities for teaching demonstrations, as well as for independent research in chemistry, and, therefore, a new laboratory was built. This laboratory became the first place in Cambridge where practical experimental tuition was part of the training of students (Dampier, 2004). There were also some meagre university facilities assigned to the chair of chemistry: an office, which was to be shared with the Professors of Botany and the Jacksonian Professor, and two small empty rooms. These rooms and an extension to them provided a small chemistry laboratory for the purpose of demonstration lectures and research; but the number of students and the amount of research increased without any corresponding growth in the facilities. This was the reason why Liveing preferred to use the laboratory of St John's for teaching purposes, in the first years of his tenure, while he kept the university facilities for his own research (Haley, 2002).

Liveing is a relatively unknown figure in the history of science. The impact of his work in Cambridge has more to do with the large amount of administrative work that he did than with his research. From his point of view, chemistry was the central experimental science and physics was the model towards which chemistry had to be directed. To give an example, in 1874 he wrote in the Student's Guide to the University of Cambridge that 'chemistry teaches laws of nature which are universal, and which find their applications whenever the structure of natural objects is under consideration'. In his Presidential Address to the chemistry section of the British Association for the Advancement of Science in 1882, he claimed that chemistry was not only a descriptive science but was about to become a predictive science based on mechanical principles, as adult sciences were supposed to be from a Whewellian point of view (Roberts, 1989, pp. 166–7). The most relevant advance in chemistry in recent years, said Liveing, 'was in the attempt to place the dynamics of chemistry on a satisfactory basis, to render an account of the various phenomena of chemical action on the same mechanical principles as are acknowledged to be true in other branches of physics'. And he added that 'I cannot say that chemistry can yet be reckoned amongst what are called the exact sciences …, but that some noteworthy advances have in recent years been made, which seem to bring such a solution of chemical problems more nearly within our reach' (Liveing, 1883, p. 479). This optimistic statement was followed by a fragment of the Philosophy of the Inductive Sciences, in which Whewell regretted that the chemical attraction known as affinity had not been reduced to the mechanical laws of attraction. Liveing thought that this moment was close at hand, especially if chemists considered the principle of conservation of energy, which was admitted both in physics and chemistry, as the cornerstone in the explanation of this property. Later in this discourse, Liveing regretted that the basic training in chemistry was still far from embodying this new approach:

> We still find chemical combinations described as if they were statical phenomena … We still find change of valency described as a suppression of 'bonds of affinity'… We still find saturated compounds spoken of as if the stability of a compound were independent of circumstances, and chemical combination no function of temperature and pressure … They present something easily grasped by the infant

mind, and schoolmasters are fond of them, but only those who
have each year to combat a fresh crop of misconceptions, and false
mechanical notions engendered by them, can be aware how much
they hinder, I won't say the advance, but the spread of real chemical
science (Liveing, 1883, p. 480).

The rhetoric of this extract, with its allusions to childish and adult science,
perhaps shows the influence of Whewell's philosophy, but more significantly,
this report helps us to link his ideas to those of the young Thomson. Indeed,
Liveing recognized that the vortex theory was not one of those 'false mechanical notions' but that it provided a 'standing ground' for the fundamental
explanation of matter (Liveing, 1883, p. 481). It is significant that, at the time
when Liveing was preparing his speech, Thomson was working on the dynamics of vortex rings, with which he tried to bring about an invasion of the field
of chemistry by the 'adult' science of physics.

In 1885, just one year after J. J. Thomson's appointment, Liveing summarized
his speculations in his only book, *Chemical Equilibrium, the Result of the Dissipation
of Energy*, in which he set out his ideas in favour of basing chemical phenomena
on mechanical principles. In the preface to the book, he manifested an even
stronger support for the vortex theory of atoms on the grounds that it gave
'more definite ideas of the manner in which dissipation of energy results in
equilibrium', and because it helped the acceptance that chemical action could
be 'founded on sound mechanical principles' (Liveing, 1885, preface). This step
forward towards a foundation on 'sound mechanical principles' meant that
chemistry could not be satisfied with the concept of affinity, 'whatever that
may mean', but that it needed to consider chemical bonding as consisting 'in a
harmony of the motions of the combined atoms in virtue of which they move
and vibrate together, and that such harmony is brought about by the general
force in nature which compels to an equal distribution of energy throughout
the universe' (Liveing, 1885, p. 83).

Liveing further advocated the introduction of mechanical principles in
chemistry using the following analogy: 'The view here advanced ... of chemical
elements is that they need not differ in substance one from another but only
in the magnitude and the form of structure of their atoms'. And he gave the
following examples: 'Hydrogen and oxygen may be compared to different bells
which though made of the same metal ring out different tones. The possibility
of the chemical combination of two atoms will depend on the possibility of
complete harmony of their motions' (Liveing, 1885, p. 89). It is interesting to
note that Liveing is being fully consistent with Thomson's views on the pre-
eminence of kinetic energy, which had been the core of his Trinity fellowship

dissertation and the central idea of his 1888 book *Applications of Dynamics to Physics and Chemistry*. Even though there is no extant correspondence between Liveing and Thomson, it is quite likely that they discussed such ideas in their meetings in Cambridge.[2] Nevertheless, in 1891, there were complaints that in the Department of Chemistry 'there is too much of the Physical Side and very little Organic teaching' (Dewar to Alexander, in Roberts, 1989, p. 171), a defect that some thought had to be corrected. The symbiotic relationship that existed between Liveing and Thomson was not exciting for everyone in Cambridge.

Pattison Muir, the incumbent chemist at Gonville and Caius College, Cambridge, is also relevant in this chapter. It is not surprising that he came to Thomson's aid with chemical and experimental advice after Threlfall left Cambridge, since they knew each other very well, and Thomson may have had particular confidence in him. Pattison Muir had been a demonstrator and assistant lecturer Owens College, Manchester, from 1873 to 1877, while Thomson was a student there. In his 1888 book, Thomson quoted profusely from Pattison Muir's book, a *Treatise on the Principles of Chemistry* (1884). The latter is also a good example of the reductionist approach to chemistry of some Cambridge scientists, which was consistent with Thomson's mindset. In close parallel to Liveing's statements, Pattison Muir declared his enthusiasm for the recent progress of chemistry: 'of late many chemists have resumed the investigation of the general conditions of chemical action, and have obtained results which give good grounds for hoping that this study ... will lead to the establishment of chemistry as a branch of the science of dynamics' (Pattison Muir, 1884, pp. 3–4).

Neither Thomson, nor Liveing, nor Pattison Muir thought that chemistry was about to disappear. A grown-up chemistry could be subsumed into a general category of physical sciences, similar to what had happened with electricity and heat. Pattison Muir tried to set the boundaries between physics and chemistry by saying that 'Chemistry deals with those reactions between bodies wherein profound modifications in the properties of the bodies occur' and by stating that 'Chemistry furnishes problems for the solution of which physical and dynamical methods are applicable. Chemical science is ever tending toward abstract truths, i.e., truths involved in many phenomena although actually seen in none' (Patison Muir, 1884, p. 4). Thus, chemistry had its own methods and aims, as far as empirical data were concerned; but, in terms of explanations, physics was the model. Moreover, the approach of chemistry and physics needed to take place from both sides. Not only did chemists need to approach

---

[2] The correspondence between Thomson and Threlfall points to the natural exchange of ideas and advice between Thomson and Liveing.

physics in search of explanations, but also physicists were expected to broaden the scope of their explanatory methods to include the realm of chemical phenomena. As an example, Pattison Muir talked about the molecular hypothesis as 'one of the lines along which dynamical science pursues its advance into the sphere of chemistry. The study of chemical phenomena is also brought within the pale of dynamical methods by the application to these phenomena of the general principles of the conservation and degradation of energy' (p. 5).

A third Cambridge chemist to bring into the picture is James Dewar. He was Jacksonian Professor of Natural Experimental Philosophy, a chair that had been created in 1783 and that had been held, at times, by people working in engineering, as well as in other applied sciences such as medicine, metallurgy, chemistry, and mechanics. Dewar was educated in Edinburgh and had his first appointments there, but he was elected Jacksonian Professor in 1875. However, he decided to make this position compatible with the Fullerian Professorship of Chemistry at the Royal Institution of Great Britain, to which he was appointed two years later. This meant that he spent more time in London than in Cambridge, since he had better facilities for his experiments there (Ross, 2004). That Dewar was suitable for a position as a Professor of Experimental Science in Cambridge can be inferred, among other factors, from the report that Maxwell wrote about him in 1874 for a position at the University of Glasgow. It is likely that, only one year later, Maxwell would have used similar arguments:

> I consider it of great importance in the present state of science that the Chairs of Chemistry in our universities should be filled by men who are able not only to extend our knowledge of the combinations of matter, but also to take part in working out the right views of combination of bodies, and who must therefore possess a thorough knowledge of physical as well as of chemical theories, and a mastery of the most accurate methods of research. The researches of Mr. Dewar in physical chemistry relate to properties of bodies which are among the most fundamental of those lying within our present range of observation, his methods of experimentation sound and well devised, and the largeness of the field which he has already exploited afford every reason to expect that he will continue to make important contributions both to the extension of chemical science and to its establishment on a firm physical basis (Maxwell, in RI, Box D/II/C/16).

Liveing and Dewar started a joint project on spectroscopic analysis on which they published many papers from 1878 onwards. For that purpose, Liveing bought a spectroscope, paid for at his own expense, since it was difficult to obtain funds for scientific research from the university. Spectroscopy could, in retrospect, be considered as a step towards physical chemistry, since it deals with the chemical composition of elements on the basis of their physical behaviour, even though it has been argued that photochemistry did not play an important role in the configuration of the new discipline of physical chemistry at the turn of the century (Dolby, 1976).

While the facilities for experimental physics, i.e., the Cavendish Laboratory, were very good, chemistry lacked a centralized laboratory for research of a high standard. On the other hand, there was more of an experimental tradition in the area of chemistry than there was in physics. The relationship between the Cavendish and the chemistry laboratories can be thought of as one of mutual help, in which neither was superior to the other. An example of this is the assistance given to Thomson by Pattison Muir, who provided, not only advice but also material help, such as the standardized solutions that Thomson needed for his experiments on ozone. Another specific example is Thomson's collaboration with Ebenezer Everett, who became the Cavendish glass blower in 1887, after training in the Chemistry Department. In fact, Liveing helped personally in this move (Thomson to Threlfall, 20 March 1887 and 4 September 1887, CUL, Add. 7654, T16 and T20). The task of blowing glass was crucial for the kind of experiments that Thomson was performing on the discharge of gases in tubes, and Everett proved to be very successful at this job, as Thomson always acknowledged.

In 1888 a new building for chemistry was built close to the Cavendish Laboratory. The new facilities consolidated the institutionalization of chemistry in Cambridge, the recovery of its independence from the Cavendish, and a shift towards research in more practical areas of chemistry. The superiority of physics in terms of facilities disappeared and, with this, Thomson's ideal of having chemistry subordinated to physics was jeopardized, although not totally abandoned. In 1894, he used a metaphor according to which physics, chemistry, and engineering appeared to be working on a common project. In his words, 'the work of chemists and physicists may be compared to that of two sets of engineers boring a tunnel from opposite ends – they have not yet met, but they have got so near that they can hear the sounds of each other's advances' (Thomson, 1894a, p. 493). This quotation brings us to a third element: the role of engineering in shaping the structure of the university.

2.8   A new tripos for engineering

Thomson needed the help of experimental chemists in order to develop his research as Cavendish Professor, and this necessity helped forge strong relationships between him and people from the Department of Chemistry. These relationships were not, however, merely contingent, but were also a manifestation of a common philosophical goal: to make use of the explanatory power of physics to account for chemical phenomena. A final example of Thomson's efforts to make physics – or, perhaps better, dynamics – central in the physical sciences is his efforts, in the late 1880s and early 1890s, to found a new tripos that would emphasize even more this centrality of physics in the organization of the sciences.

In the 1830s and 1840s, several British universities created chairs of engineering to cater for the demand for engineers in an increasingly industrialized country. King's College London, was the first university to have a professor of engineering, after which Glasgow, University College London, Trinity College, Dublin, and Queen's College, Belfast, followed suit. In Cambridge, discussions on creating a School of Engineering started in 1845, but that project saw opposition from the most conservative dons of the university, for whom a School of Engineering was tantamount to having a warehouse giving university degrees. The little engineering one could find in the Cambridge syllabus was traditionally taught under the name of mechanics by the Jacksonian Professor of Natural Experimental Philosophy.

As we noted earlier, in 1875, a chemist, Dewar, was appointed to the Jacksonian professorship and a new chair for mechanism and applied mechanics was created. The new professor was to teach '(1) The Principles of Mechanism, (2) The Theory of Structures, (3) The Theory of Machines, and (4) The Steam Engine and other Prime Movers' (Hilken, 1967, p. 30), topics in which a Cambridge student could take a special examination. The new professor was also expected to play a central role in the eventual creation of a School of Engineering in the university, but such a school was proving elusive. James Stuart, the first Professor of Mechanism, worked on two fronts in order to consolidate his department. On the one hand, he was keen to obtain a physical space for engineering, for which the University provided an old wooden hut next to the Cavendish; on the other hand, he was eager to establish a new tripos, different from both theMathematical and Natural Science Triposes, which would give students of mechanism a greater prestige than the current 'special examination'.

The discussions leading to the first proposal for the tripos reveal that the reasons Thomson and others were supporting Stewart had to do with the gap between theoretical and experimental science in Cambridge. The two existing triposes appeared to be so clearly disconnected that both Liveing and Thomson thought it advisable to create a tripos with a broad content of both mathematics and practical science. The proposal regretted 'the want of mathematical training in the great number of engineering students, and it was most desirable that the mathematical training should be considerably increased; at the same time these men could not possibly devote three or even two years exclusively to mathematics'. Actually, 'the proposal was quite as much for students of Physics as for engineering students. The students of Physics wanted a great deal of Mathematics but a very different kind of Mathematics from that in Part I. of the [Mathematical] Tripos' (CUR, 1886, p. 405).

Later in the discussions, Thomson made it clear that the reason he was supporting the new tripos was not so much for the sake of engineering but because of the lack of mathematical content in the Natural Science Tripos. That is why he agreed with the suggestion that 'if the Mathematical Tripos could be altered that would be the best arrangement'. But this would take too long and 'the need of mathematical training for these students was very urgent' (CUR, 1886, p. 407). What was happening was that, due to the difficulty and the apparent uselessness of much of the mathematics that was taught in the Mathematical Tripos, students in the Natural Science Tripos did not feel encouraged to attend the Mathematical Tripos lectures.

Shaw, echoing Thomson's interest, put it very clearly in the Senate House when he said that 'to those observing what students of physics, chemistry, and engineering in the University actually learnt, it became at once obvious that there was a good deal of ground common to all, and that the examination which covered the common ground and included certain portions of Mathematics would be extremely useful' (CUR, 1887, p. 77). In his conversations with other engineers, Thomson had also drawn the conclusion that 'a knowledge of Physics was absolutely necessary for a scientific training for engineers; indeed one of the most successful Professors said that he thought it more necessary even than practical engineering, for a student could get practical engineering after he had left the University, but he would have no other opportunity of making the simple physical experiments contemplated by the regulations' (p. 75). This helps us to understand why Thomson wanted physics to be the central discipline between the too theoretical Mathematical Tripos, and the too experimental Natural Science Tripos in which physics played only a subsidiary role.

It is quite significant that the final proposal for a Tripos for Mechanics, Physics and Chemistry was signed by Liveing and Dewar, but did not have the approval of Thomson and Glazebrook: physics was considered an optional subject, not the centre of the tripos. This proposal was, nevertheless, turned down, and Stuart resigned in 1889. A new professor was appointed on 12 November 1890. James Alfred Ewing was already Professor of Engineering and Drawing at University College, Dundee, and this experience gave him the authority to work successfully towards the creation of a Department of Engineering in Cambridge.

After the failed attempt to use the Engineering Tripos as a way to consolidate physics as the queen of sciences, Thomson and Glazebrook argued that the optimal situation, as far as physics was concerned, would be that of a tripos in which students could receive almost all the mathematical training acquired in the Mathematical Tripos together with the physics and chemistry in the Natural Science Tripos, without the need of doing the complete training of the first years of the Mathematical Tripos. In a report submitted to the Senate House of the university in 1890, they advocated a 'mixed Tripos of Mathematics and Physics' (CUR, 1890, pp. 558–61). In this proposal, physics – and not chemistry – would have been compulsory. This time the proposal was signed by Thomson, but not by the Professors of Chemistry, and it was not approved by the university. Also in the academic year 1891/92, discussions were held in order to split the examination in physics and chemistry into two separate and independent examinations in the Natural Science Tripos. The curriculum in chemistry was to include a voluntary exercise in fundamental physics, mainly electricity, and the curriculum in physics was also to include a voluntary exercise in chemistry. These discussions point to a further separation between these disciplines from a pedagogical and institutional point of view.

In 1894, the new Engineering Tripos was finally created, and Ewing managed to convince the university to build a new building for his Department on the recently purchased site between the Cavendish and the Chemistry Department. He thought it 'exceedingly desirable that if, or rather when, such buildings are erected they should be close to the Cavendish and Chemical Laboratories, for much of the work of students of engineering would be done in those places, and proximity of the buildings would greatly facilitate the arrangement of lecture hours, &c' (CUR, 1891, p. 563). Far from achieving Thomson's dream of establishing physics as the queen of the physical sciences, the implementation of the new tripos finally consolidated the split and subsequent specialization of physics, chemistry, and engineering. In the definitive organisation of the sciences in Cambridge, specialisation won out over Thomson's dream of unification of the 'physical sciences'.

3

The ether and the corpuscle: from waves to particles

3.1   Electric discharge in tubes

Which comes first, theory or experiment? And, in any case, what is theory? In the prologue to his *Notes on Recent Researches on Electricity and Magnetism*, published in 1893 as a sequel to Maxwell's *Treatise*, J. J. addressed these questions with an eye on Cambridge Mathematical Tripos wranglers, 'who have a great tendency to regard the whole of Maxwell's theory as a matter of the solution of certain differential equations, and to dispense with any attempt to form themselves a mental picture of the physical processes which accompany the phenomena they are investigating' (Thomson, 1893, p. v). In less than ten years as Cavendish Professor, J. J. had come across many Cambridge graduates who, like himself over a decade earlier, could barely perform a single experiment after more than three years of an exclusively mathematical training, and who were largely oblivious to the relationship between their mathematics and the phenomena in the laboratory. Research was, for such graduates, nothing more than the task of solving more complex mathematical problems based on already existing theories and formulae. This could have, in the long term, hindered the advance of science for lack of creative ideas to guide new avenues of experimental research: 'I think that this state of things is to be regretted, since it retards the progress of the science of Electricity and diminishes the value of the mental training afforded by the study of that science' (p. vi). And, with a hint of humour, he added:

> The use of a physical theory will help to correct the tendency – which I think all who have had occasion to examine in Mathematical Physics will admit is by no means uncommon – to look on analytical processes as the modern equivalents of the Philosopher's Machine in

the Grand Academy of Lagado, and to regard as the normal process
of investigation in this subject the manipulation of a large number
of symbols in the hope that every now and then some valuable result
may happen to drop out (Thomson, 1893, p. vi).

With his emphasis on mental models, or, as he now called them, the 'geometrical and physical method', J. J. Thomson was an advocate for the long standing
Victorian tradition that identified rationality with the production of mental
mechanical models to explain phenomena, which was regarded as the final
triumph of Newtonian mechanics: the paradigm of adult science of which we
spoke in the previous chapter. One of the big epistemological questions was,
of course, that of the actual reality of such mental models: were they purely
heuristic tools or real mechanisms in nature? Generally speaking, Victorian
physicists tended to favour the former, but admitting that even if the *actual*
mechanism invoked in each particular case might be wrong, there had to be
some mechanical process behind every physical phenomenon. Along these
lines, Thomson had argued for dynamical models in an earlier paper, saying
that 'in the first place there is the mental satisfaction to be got by explaining things on dynamical principles; and again, there is the certainty that the
method is capable of completely solving the question', after which he added
in brackets the cautious statement 'whether we can make it do so is another
matter' (Thomson, 1887, p. 473).

In the preface of *Notes on Recent Researches in Electricity and Magnetism*,
J. J. acknowledged the dangers of taking models too seriously, since extrapola-
tions from them could lead to conclusions not justified by the corresponding
analytical theory. This, however, should not be a problem 'if we remember
that the object of such theories is suggestion and not demonstration. Either
Experiment or rigorous Analysis must always be the final Court of Appeal; it
is the province of these physical theories to supply cases to be tried in such a
court' (Thomson, 1893, p. vii). Mental models were, at least in this preface, a
highly recommended tool for suggesting new experiments and understanding
yet unexplained phenomena, as well as a way to further develop or polish a
given mathematical theory.

But did J. J. Thomson himself work along these lines? The answer to this
question is not straightforward. As Falconer (1985) thoroughly documented,
the relationship between theory and experiment in Thomson's work was
always very loose, since he cared very little for quantitative precision and close
correspondence with experimental results. He was more interested in results
that were *suggestive*, and from which he could extrapolate and speculate on
new theories. In the 1880s, he was speculating on the nature of matter and

extrapolating, from the analysis of vortex rings, an overarching theory of everything, without much reference to experimental results. As Cavendish Professor, he started to experiment on electrical discharges in gases in what looks like a haphazard collection of data on these phenomena; and that was partly the case. Convinced, as Maxwell was, that electric discharge in tubes was a most important phenomenon, not so much due to 'the beauty of the experiments, as to the wide-spread conviction that there is perhaps no other branch of physics which affords us so promising an opportunity of penetrating the secret of electricity' (Thomson, 1893, p. 53), J. J. had first to familiarize himself with the methods and techniques involved in such experiments, as well as justifying the academic validity of this field of research. Actually, reference to the beauty of discharge tubes is not accidental: it was quite common in Victorian popular science to treat the colours and shapes of the light obtained in experiments of electrical discharge in gases as a source of entertainment in science lectures for popular audiences. Thomson's first challenge was to legitimize an experimental field that, in words of Arthur Schuster some decades later, 'had better be left to be played by cranks and visionaries' (Schuster, 1911, p. 52). That is partly why the second chapter of *Recent Researches* was wholly devoted to a qualitative survey of all the phenomena known about electric discharge through gases, a topic on which 'there is no summary in English text books' (Thomson, 1893, p. 53).

The link between theory and experiment was loose, in the sense that Thomson used his (many) qualitative observations and (fewer) quantitative results to illustrate a theory of dissociation based on his vortex ring theory of the atom. Until about 1890, his work on electric discharge in tubes played a crucial role neither in shaping his theories on electricity and matter, nor in the design of experiments to prove or disprove his models. His main interest, both in his pre-1890 theories and in his experiments, was to discuss electric discharge in gases in terms of the dissipation of electric energy. This was in natural congruence with his overall aim of uniting physics and chemistry through the application of dynamics. Only after 1890 did his attention turn towards the exploration of the transfer of electric charge, a change that came hand in hand with his abandonment of the vortex ring theory of the atom.

The first instalment in a long series of speculations to understand electric discharge appeared in the *Philosophical Magazine* as early as 1883, a year before he was elected director of the Cavendish Laboratory, which shows that his interest in the topic preceded his appointment. Certainly, his new job triggered a long series of experiments, which, had he remained *only* a mathematical physicist, he might never have had the chance to perform (de Solla Price, 1957). The fundamental elements of this 1883 theory, which remained basically unaltered until around 1890, were based on his vortex ring model of the atom and on the Clausius and Williamson dissociation theory, 'according to which the molecules of a compound gas are supposed not to always consist of the same atoms of the elementary gases, but that these atoms are continually changing partners' (Thomson, 1883c, p. 428). Thomson proved mathematically that, under certain circumstances, two vortex rings travelling in the same direction could eventually form a mechanically stable system. In that case, both rings would unite and travel together. In his usual speculative style, he went on to suppose that 'the union or pairing in this way of two vortex rings of different kinds is what takes place when two elements of which these vortex rings are atoms combine chemically' (p. 427). We should note that Thomson was here considering two vortex rings as the components of the gas molecules, even in those cases where the gas was known to be composed of three or more elements. Thus, one vortex ring could represent one atom, or more than one atom, or, as we saw in the previous chapter, one atom could be represented by a number of vortex rings.

Thomson then introduced the concept of the ratio between paired and free time, i.e., the ratio between the time vortex rings were bound together forming a molecule, and the time they were on their own. This ratio would depend on the temperature and the pressure of the gas. When dissociated, a gas no longer exhibited the properties of the chemical compound, but rather of its constituent elements, and this would happen when the ratio between paired and free time was small enough. The presence of an electric field, which in the Maxwellian paradigm was a tension along the lines of force in the ether, would be responsible for an increase in the energy in the gas with effects analogous to those observed with a rise of temperature: at first, this increase in energy would translate into more mechanical energy of the compound molecules without any chemical change (i.e., without dissociation); but, as the energy kept rising, the ratio between paired and free time of the molecules would reach a threshold beyond which the gas dissociated. The process of dissociation would absorb much of the energy, thus discharging the electric field, and the later recombination would emit energy in the form of heat, with the characteristic light patterns of this kind of discharge. In his words,

> The disturbance to which the gas in an electric field is subjected
> makes the molecules break up sooner into atoms than they otherwise
> would do, and thus diminishes the ratio of the paired to the free
> times of the atoms of the gas; as the intensity of the electric field
> increases, the disturbance in some places may become so violent that
> in these regions the ratio of the paired to the free times approaches
> the value it has when the gas is about to be dissociated. At this

point any diminution of this ratio consequent upon an increase in
the intensity of the field will absorb a large amount of energy; this
energy must come from the electric field; and we should thus get the
phenomenon of the electric discharge (Thomson, 1883c, p. 431).

Following the argument of the previous chapter, it is worthwhile noting that
this discharge mechanism was in tune with Thomson's interest in bridging the
gap between physics and chemistry, and in trying to develop a chemical theory
based on dynamics.

From this model of discharge, Thomson inferred that one experimental magnitude central to his theory and worth measuring was the 'electric strength
of gases', i.e., the maximum potential a gas could support without discharge,
thinking that this would turn out to be a characteristic quantity for each gas.
But his experiments up to 1890 showed that the electric strength was far from
being a fundamental of gases, since it depended on other factors related to the
electrodes, such as their material, kind of surface, shape, size, and separation.

In 1886, Thomson extended his vortex ring dissociation theory to also explain
the asymmetry of the discharge the positive and the negative poles. One of the
basic visual features of the electric discharge in rarefied gases is that the pattern is clearly different at the two extremes of the tube. The region immediately adjacent to the cathode (called the 'Crookes space' or 'dark space') varied
in length depending on the conditions inside the tube, but not on the material from which the cathode was made. Next came the negative glow, another
dark space known as the Faraday space, and finally a collection of striations. In
a totally qualitative and highly speculative paper (Thomson, 1886), Thomson
started from the consideration that a gas molecule was composed of two similar vortex rings on parallel planes and with their centres on the same axis. In
the presence of an electric field, the molecule would move in the direction of
the lines of force or in the opposite direction. Using hydrodynamic considerations, J. J. argued that the tension in the lines of force in the ether increased
the radii of the vortex rings. Thus, the two rings forming a molecule moving in
the direction of the lines of force would have different diameters. An increase
in the radius of one ring would, of course, affect its speed and, thus, the two
rings would move apart from each other if the molecule was moving in the direction of the field (or closer together, if moving in the opposite direction).

With this in mind, J. J. suggested that molecules of this kind would have a
greater tendency to split closer to the negative than to the positive electrode,
which, in turn, would explain the darkness in the 'Crookes space'. The negative glow would be the consequence of the recombination of the vortex rings
into molecules, a light-emitting process. Beyond this, the intensity of the field

would no longer be high enough to give rise to any further molecular splitting, thus explaining the dark interval. To account for the pattern close to the positive electrode, J. J. fixed his attention on the molecules moving away from the positive electrode. The phenomenon was qualitatively the same as in the proximity of the negative electrode, except for the fact that, following the hydrodynamic implications of his model, molecules had less tendency to split close to the positive electrode than to the negative.

This might sound like an extremely speculative and ad hoc model, but we should not forget that the problem of asymmetry between the positive and negative electrodes was a constant source of headache for all scientists working in this field, as well as an unending source of inspiration for the Victorian scientist given to speculation and modelling. The problem would only be eventually solved by the corpuscle-electron and the essential asymmetry between positive and negatively electrified particles.

3.2   From discharge tubes to Faraday tubes

Recent Researches contains a comprehensive up-to-date survey of experiments and theories related to the latest advances in electromagnetism since the publication of Maxwell's *Treatise*. As we just saw, the second chapter included all the experimental progress in the field of electric discharge in gases. But by 1893, when he published this volume, J. J. had already abandoned his vortex ring theory. The two most relevant new ideas in the book were his new interpretation based on Faraday tubes and Hertz's experiments proving the existence of electromagnetic waves. The former form the core of the first chapter, while the latter appear only in Chapter 5. This may lead the reader to the wrong conclusion that Hertz's experiments were unrelated to the formulation of the theory of Faraday tubes.

In 1888, Hertz obtained indisputable evidence for the existence of electromagnetic waves and, indirectly, for the existence of the ether, as opposed to the typically Continental idea of action-at-a-distance. The reception of Hertz's experiments in Britain was more than enthusiastic, since they seemed to finally prove the validity of Maxwell's framework (Hunt, 1991, pp. 158–62). J. J. Thomson was one of many British physicists to replicate the experiments and to regard them as the *experimentum crucis* for the Maxwellian world-view; but his work was also influenced by one particular technique that Hertz had developed to obtain rapidly alternating currents. As early as 1881, Thomson had longed for an experimental way to test the constancy of the specific inductive capacity at high frequencies (Falconer, 1985, p. 67). Between 1888 and 1890, after learning about Hertz's technique, he slowed down the pace of his experiments on electric discharge to test Maxwell's theory at high frequencies. His main interest was the relationship between the velocity of propagation of electrical disturbances in a conductor and that in the dielectric surrounding the conductor, since Hertz's experiments seemed to suggest that the two velocities were different, whereas, according to Maxwell, they should be the same. Thomson performed experiments not only on solids, but also in liquids and gases (the latter being his main field of expertise) and proved that there was no great difference between the two velocities, which lent further support to the Maxwellian tradition (Thomson, 1889a, c, d). But in doing so, he also showed that the velocity of discharge in tubes, a magnitude that had so far been neglected in his vortex ring theory, was unexpectedly close to the speed of light. The inevitable consequence was his progressive abandonment of the theory based on vortex rings.

Interestingly, however, J. J. did not present the transformation of his theory as a dramatic change in his views, but underlined the continuity between the old and new theories by emphasizing what remained unchanged. In 1890, he wrote that the new results were 'in accordance with the view of the electric discharge through gases which I gave in the *Philosophical Magazine* for June 1883', the basic premise of which was that 'the provision of a supply of atoms by the splitting up of the molecules is the essential accompaniment of the electric discharge through gases' (Thomson, 1890a, p. 359). Consistent with his epistemological considerations on the role of models, J. J. apparently abandoned the vortex ring theory without much trauma, partly because he was keeping some of the basic elements that it contained, specifically the dissociation of molecules into atoms. The great change, however, was that this dissociation no longer took place on the basis of dynamical considerations, but on electrical ones:

> We may regard the atoms in the molecule as being in oppositely polarized states; one atom behaving as if it were charged with a quantity of positive electricity, the other as if it were charged with an equal quantity of negative. When the atoms are together in the molecule they neutralize each other's action at points outside the molecule, which behaves as if it were electrically neutral; but as soon as the atoms separate, since each one is essentially polarized, the gas acquires energetic electrical properties, and by the motion of its atoms electricity can be carried from one part of the gas to another. The ease with which the gas can be made to conduct electricity depends upon the ease with which its molecules can be split up into atoms (Thomson, 1890a, p. 359).

This move from directly dynamic considerations towards electric arguments was also partly motivated by J. J.'s renewed interest in electrolysis: the similarities between it and discharge in tubes were too strong to be neglected, even though the common explanations of electrolysis based on the existence of molecular charges were not well regarded by Maxwellians.[1] At the same time, however, Thomson's experiments on the rate of propagation of electric discharge, which showed it to be close to the speed of light, meant that the motion of charged atoms or molecules was simply not enough to account for this high speed, since charged atoms were not energetic enough: 'The very rapid rate with which the electric discharge is propagated through a rare gas compels us to admit that the electricity is not carried by charged atoms moving with this velocity' (Thomson, 1890c, p. 132). That is why he needed to introduce some structure in the dielectric that would act, like a wire in the case of electricity in solids, as a channel for conduction. Thomson was trying to combine in one model the discreteness of electrification with the continuity of the medium. With this in mind, he first introduced a mechanism based on Grotthus chains, and, soon afterwards, his Faraday tubes of force.

The former he presented in the following terms:

> Before the electric field is intense enough to cause discharge, the induction in the field polarizes the gas. We may regard this polarization as being equivalent to the formation of chains of molecules analogous to the 'Grotthus chain' in electrolysis. As the intensity of the field increases, suppose the molecules in one of these chains near an electrode, say the negative, interchange their atoms; and that *it is not merely those molecules which are next the electrode which split up, but that the decomposition of the molecules extends along an appreciable length of the chain*. The positively electrified atoms will cling to the negative electrode, and after a time, depending upon the number of free atoms, the distance between them, and their mutual attractions, the chain will resume its original molecular condition (Thomson, 1890c, p. 132, emphasis in the original).

Grotthus chains were an old mechanism conceived by Theodore von Grotthus and by Humphry Davy at the beginning of nineteenth century, and reintroduced by Faraday in 1833, to account for the transmission of electricity in electrolysis (Darrigol, 2000, p. 81). J. J. was suggesting that, in a tube filled with a

---

[1] In the *Treatise*, Maxwell (1873) had written: 'This theory of molecular charges may serve as a method by which we may remember a good many facts about electrolysis. It is extremely improbable however that when we come to understand the true nature of electrolysis we shall retain in any form the theory of molecular charges, for then we shall obtained a secure basis on which to form a true theory of electric currents, and so become independent of these provisional theories' (§269).

rarefied gas and in the presence of an electric force, the continuous association and dissociation of molecules into two oppositely charged ions was guided by the electric force. From his point of view, there was observational evidence for this mechanism, since 'this breaking-up of the current into a series of separate pieces shows itself in the stratifications observed when the discharge passes through a gas at low pressure' (Thomson, 1890c, p. 133). The Grotthus chain thus kept the association and dissociation of molecules that had been present in his vortex-ring theory while, at the same time, introduced the quantification of electric charge that was common in the explanations of electrolysis.

The final change on his views on electricity occurred only a few months later, when J. J. presented his idea of Faraday tubes, an idea that not only helped one visualize the arrangement of Grotthus chains, but also eventually became the main structure on which to base an all-embracing theory of electrodynamics. The first public appearance of these tubes was in a paper in *Philosophical Magazine* on early 1891, and J. J. gave the following justification:

> Methods such as this, of materializing, as it were, mathematical conceptions, seem to have a use even where, as in the case of Electricity, the analytical theory is well established; for any method which enables us to form a mental picture of what goes on in the electric field has a freshness and a power of rapidly giving the main features of a phenomenon, as distinct from the details, which few can hope to derive from purely analytical methods. Experience has, I think, shown that Maxwell's conception of electric displacement is of somewhat too general a character to lend itself easily to the formation of a conception of a mechanism which would illustrate by its working the processes going on in the electric field. For this purpose the conception of tubes of electrostatic induction introduced by Faraday seems to possess many advantages. If we regard these tubes as having a real physical existence, we may, as I shall endeavour to show, explain the various electrical processes, – such as the passage of electricity through metals, liquids, or gases, the production of a current, magnetic force, the induction of currents, and so on, – as arising from the contraction or elongation of such tubes and their motion through the electric field (Thomson, 1891, pp. 149–50).

This long statement deserves some attention since it summarizes one of the problems that plagued Faraday tubes throughout their existence in the mind and work of J. J. Thomson. The tubes were first introduced as a visual tool that might help explain phenomena and guide research on something that was, in principle, already fully explained by Maxwell's notion of electric displacement.

Here, the heuristic character of Faraday tubes seems obvious. But later on, he tells us that we have to think of Faraday tubes as having a 'real, physical existence', and not only as mental tools for suggesting further research. As we shall see, many of the properties of electricity can be ascribed to the mechanical and dynamical properties of the Faraday tubes. The different layers of reality that J. J. used in the vortex ring theory were equally present here. In the former, atoms were real, but they were, at a more fundamental level, epiphenomena of the ether. Faraday tubes were also structures of the ether and, thus, had the 'same, real existence' as the ether itself. Furthermore, as this long quotation shows, J. J. Thomson did not introduce the Faraday tubes as a tool to fill a gap in one particular explanation, but as the basic structure of the ether on which to base the understanding of all electromagnetic phenomena. This is particularly clear in the structure of *Recent Researches*. The book was designed as a step forward from Maxwell's *Treatise* and took Faraday tubes as the new, alternative, foundation stone on which to base further developments in the science of electricity. That is why the first chapter was devoted exclusively to introducing the reader to the properties of the tubes and their use in describing all electrical and magnetic phenomena.

But, what are Faraday tubes exactly? Electrostatically, they are unit tubes of electrostatic induction, all with the same strength corresponding to the electrolytic unit of charge. Mechanically, they are structures in the ether in the form of vortical tubes that begin and terminate in matter or form closed circuits. These tubes have a direction and the atoms on which they begin and end receive a unit of positive and negative electrification respectively. To support his new theoretical device, J. J. stressed that these structures were already present in the work of Faraday and Maxwell, at least in the first sense of connecting positive and negatively electrified bodies (without the possibility of closed loops). In doing this, he portrayed his ideas as a continuation of the work of the two main British authorities on electricity and magnetism. In fact, J. J. in the *Treatise* explicitly quoted Maxwell's description of how to generate a tube of induction force from a line of force: 'If the line of force moves so that its beginning traces a closed curve on the positive surface, its end will trace a corresponding closed curve on the negative surface, and the line of force itself will generate a tubular surface called a tube of induction' (Maxwell, 1873, §82). But, as Darrigol pointed out, Thomson's Faraday tubes were a complex hybrid of concepts from Faraday and Maxwell, as well as from William Hicks, Poynting, Helmholtz and Schuster, whom J. J. characteristically did not mention at all (Darrigol, 2000, p. 269).

As with the vortex ring theory, Faraday tubes were structures emerging from the dynamics of the ether. Even though, in the first instance, J. J. did not speculate on the way the tubes were constituted, the analogies with vortical movements were ubiquitous. Thus, at the beginning of *Recent Researches* he said that 'The property of the Faraday tubes of always forming closed circuits or else having their ends on atoms may be illustrated by the similar property possessed by tubes of vortex motion in a frictionless fluid, these tubes either form closed circuits or have their ends on the boundary of the liquid in which the vortex motion takes place' (Thomson, 1893, pp. 3–4). And, at the end of the first chapter, he emphasized the same idea, saying that 'the analogies which exist between their properties and those of tubes of vortex motion irresistibly suggest that we should look at rotatory motion in the ether for their explanation' (p. 52).

The rotating movement of the ether in and around Faraday tubes explained electric and magnetic phenomena along the lines of the general framework already set out in *Applications of Dynamics to Physics and Chemistry* (Thomson, 1888): the kinetic energy due to these vortical movements constituted the potential energy of the electrostatic field, 'while when the tubes themselves are in [translatory] motion we have super-added to this another distribution of velocity whose energy constitutes that of the magnetic field' (Thomson, 1893, p. 5). And he developed the mathematics of these assumptions, introducing the notion of polarization which was 'mathematically identical with Maxwell's "displacement"', but had 'a different interpretation' (p. 6). Since the ends of Faraday tubes had opposite electrifications, a tube had direction. Polarization at one point of a dielectric was a vector quantity that represented the density and direction of Faraday tubes of force per unit volume. Since Faraday tubes in a dielectric 'cannot be created nor destroyed', a change in polarization could take place only when tubes moved or deformed. And, from these changes in polarization, which, at the end of the day, were dynamical changes treatable with the usual methods of dynamics, Thomson showed the direct connection between his method and the energy transfers obtained using Poynting's vector.

Two atoms would form a stable molecule when united by a short tube of force, i.e., a Faraday tube of molecular dimensions. At this stage, however, he did not speculate much on the relationship between Faraday tubes and atoms of ordinary matter. In his previous vortex theory, all phenomena were supposed to be manifestations of the ethereal vortex rings and their movements in the ether. There was no asymmetry between ether and matter precisely because, in a way, matter did not exist as essentially distinct from the ether but only as a manifestation of it. Furthermore, and more importantly, there was only one kind of structure in the ether: the vortex rings. Now, on the other hand, atoms were essentially different from the Faraday tubes, since these

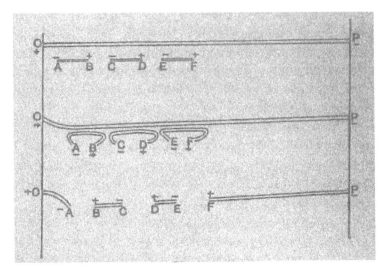

Figure 3.1 The mechanism of electric discharge in rarefied gases using Faraday tubes (Thomson, 1893, p. 46). By permission of Oxford University Press.

ended – or 'fell', as he normally said – on atoms. It was not until 1895 that he actually started to think explicitly the relationship between Faraday tubes, ether, and atoms.

But, to continue with the story of electric discharge in tubes, let us look at the role of Faraday tubes in this process. Figure 3.1, taken from *Recent Researches*, is quite self-explanatory: the short Faraday tubes AB, CD and EF represent molecules of the gas which, in the presence of a field between the ends of the tube, represented by the long tube OP, line up in the direction of the field. The molecules of the gas thus polarized attract the long tube OP, since this is of opposite sign to AB, CD and EF. When the field is strong enough, there is a splitting of the tubes, which means a splitting of the molecules, creating a Grotthus chain. In his usual style of extrapolating without much justification, J. J. assumed that a similar process happened in all media, including solid metals.

This mechanism kept the main ideas that he had developed around 1890. Discharge of electricity through gases involved molecular dissociation, but a dissociation that had its foundation in electrical considerations (Faraday tubes) and no longer in energy considerations (vortex rings). Faraday tubes had introduced a previously neglected discreteness in the electric charge, which was consistent with the science of electrolysis: following Maxwell, electric charge had to be a phenomenon arising at the interface between ether and matter, but the presence of Faraday tubes, a discrete structure of a continuous medium, allowed for discreteness in charge without contradicting the essentially continuous Maxwellian paradigm.

3.3   Tubes, electricity, and matter

The concept of Faraday tubes brought about a shift towards an atomization in the structure of electricity and, indirectly, towards the notion of discrete charge. Prior to 1890, J. J. did not think in terms of discrete charges but rather in terms of exchanges of energy. The analogy with electrolysis and the use of Faraday tubes changed things: charge became a phenomenon at the interface between ether and matter, between Faraday tubes and matter, and, since the former had a fixed and specific strength, the magnitude of electric charge was not continuous but discrete. If the tubes of force were real physical entities, and not merely ideal devices, there should be an actual physical limit to their divisibility. This idea opened the door to a quantification of energy and charge within the framework of a continuous ether: discreteness was not an essential quality of the ether, as other contemporary theories suggested. In other words, Faraday tubes allowed for a theory in which electric charge was at the same time discrete and a boundary phenomenon, not a substance. In a very programmatic statement, he wrote:

> From our point of view, this method of looking at electrical phenomena may be regarded as forming a kind of molecular theory of Electricity, the Faraday tubes taking the place of the molecules in the Kinetic Theory of Gases: the object of the method being to explain the phenomena of the electric field as due to the motion of these tubes, just as it is the object of the Kinetic Theory of Gases to explain the properties of a gas as due to the motion of its molecules. These tubes also resemble the molecules of a gas in another respect, as we regard them as incapable of destruction or creation (Thomson, 1893, p. 4).

The analogy between the structure of the electric field and the kinetic theory of gases is quite revealing since it immediately raises the question of what happens with matter: is it still to be regarded as atomic and, if so, what is the relationship between atoms and the Faraday tubes? These questions were not addressed when Faraday tubes were first suggested, and only started to creep in through the relationship between electrification and chemical action, via electrolysis, as well as through the problem of asymmetry in the discharge patterns. Because, if electrolysis showed a constancy in the charge while the patterns were clearly asymmetric in a tube, there could only be one explanation: the structure of molecules had necessarily to account for such asymmetry between positively and negatively charged atoms.

After the publication of *Recent Researches*, Thomson began a series of experiments on the electrolysis of steam and, later, a broader analysis on the electrolysis of gases, to see if he could still stand behind his idea that 'the discharge through gases is accompanied by chemical changes analogous to those which take place in electrolytes conveying currents' (Thomson, 1894b, p. 92). The results he obtained were somewhat puzzling and difficult to account for since, contrary to what was normally the case in the electrolysis of solutions, the sign of the charges of the elements in the electrolysis of gases was not always the same. In solutions, 'the persistency of the sign of the electric charge carried by an ion is almost as marked a feature as the constancy of the magnitude of the charge', while that was not the case in the electrolysis of gases: 'here an atom of hydrogen does not always carry a positive charge, nor an atom of chlorine always a negative one; each of these atoms sometimes carries a negative charge, sometimes a positive one' (Thomson, 1895a, p. 511). On first analysis, this seemed to undermine much of Thomson's idea that electrolysis and discharge were strongly related phenomena; but, after some thought, the idea that there was an asymmetry between positive and negative electrification was reinforced: even in the case of elementary diatomic gases like hydrogen, ions behaved chemically differently when electrified positively or negatively. In other words, the different behaviour at the positive and negative electrodes reflected the asymmetry between positive and negative electrification, irrespective of the substance that was being electrified (Thomson, 1895a, p. 255). Following this thread, J. J. now sensed that electrification of molecules and chemical action were two sides of the same coin and, therefore, he sought to understand this relationship better. This led him to speculate more explicitly on the relationship between charge and matter, making use of his Faraday tubes:

> The connexion between ordinary matter and the electrical charges on the atom is evidently a matter of fundamental importance, and one which must be closely related to a good many of the most chemical as well as electrical phenomena. In fact a complete explanation of this connexion would probably go a long way towards establishing a theory of the constitution of matter as well as of the mechanism of the electric field. *It seems therefore to be of interest to look on this question from as many points of view as possible, and to consider the consequences which might be expected to follow from any method of explaining, or rather illustrating, the preference which some elements show for one kind of electricity rather than the other* (Thomson, 1895b, p. 512, my emphasis).

I emphasized the last sentence because, at that time, and true to his implicit philosophy of science, J. J. did not seem to think of this particular model based on Faraday tubes as any more ontologically real than other models.

Furthermore, he was directly interested at that time not in building a model of matter, but in understanding the relationship between electrification and chemical action, or valency. While the Grotthus-chain mechanism had given him an explanation of the velocity of discharge, he now needed a theory that could account for electrification. That the new model he developed did not give a satisfactory explanation of the velocity of discharge was not important since, for him, models were only explanatory tools for particular problems, and there was no need for full consistency between models, as the above quote shows.

The contrast between the earlier vortex atom theory and the new Faraday tubes was made explicit in this 1895 heuristic model, where Thomson argued that the new theory was only a step, perhaps an important one, towards a better understanding of the nature of matter. In his words,

> Now let us consider the atoms on which these tubes end. Let us suppose that these atoms have a structure possessing similar properties to those which the atoms would possess if they contained a number of gyrostats all spinning in one way round the outwardly drawn normals to their surface. Then one of these atoms will be differently affected by a Faraday tube, and will possess different amounts of energy according as the tubes begins or ends on its surface (Thomson, 1895b, p. 513).

Two things are important here: the totally heuristic character of the rotating gyrostats, and his sticking to rotating movements as the main characteristic on which to focus. Like in the vortex ring theory, here he speaks of a certain structure within the atom. With this model, he also tried to account for the affinity of elements and the formation of molecules, which was consistent with his idea of giving chemistry the status of a science dependent on physical models. Affinity was here understood as a consequence of the way that Faraday tubes ended the atoms. The atoms were described in terms of gyroscopic structures, and the moment of momentum created by the interaction of a Faraday tube with the vorticity and the momentum of the atom explained the different affinities between different kinds of atoms. The gyroscopic structures also accounted for the electrochemical behaviour of atoms. Thus, chemical behaviour could be explained in terms of physical mechanisms and not simply as the result of electrical forces. As an example, Thomson cited the apparent asymmetry in the bond between hydrogen and chlorine: a negatively electrified hydrogen atom and a positive chlorine atom would experience less attractive force than a positive hydrogen atom and a negative chlorine atom, although the electrostatic force was the same, due to the different rotational action of the gyrostats on the also rotating Faraday tubes. This is what Thomson meant when he said that 'when charged atoms are close together, there may be forces partly electrical, partly chemical, in their origin in addition to those expressed by the ordinary laws of electrostatics' (Thomson, 1895b, p. 518).

Following the chemical side of this model, Thomson discussed a few simple cases so as to show that electrostatic attraction was not the only condition for two ions to form a chemical compound. Thus, for instance, in a solution with hydrogen and chlorine, both molecular compounds would be split into atoms, each with positive and negative electricity. With only electric arguments, all atoms would be able to form molecules of HCl, with positive Hs and negative Cls interacting as well as negative Hs with positive Cls. But in the latter case, and due to the torsion of the Faraday tube potentially uniting both atoms, there would be repulsion rather than attraction. Only when both atoms managed to exchange the sign of their electrification would this torsion cease to be repulsive and the HCl bond easily take place. This charge exchange could only take place in the presence of a field different from the one created by the two individual atoms; a field that would force the atoms close enough so as to exchange the sign of their electrification or, what amounts to the same thing, so that it forced a shift in the extremes of the Faraday tube potentially linking both atoms. From here, he inferred a general law by which 'Chemical action does not (in general) take place between a single pair of molecules alone in the field, but requires the formation of aggregates either of the interacting molecules or of some third substance which is either a conductor of electricity or has a large specific inductive capacity' (Thomson, 1895a, p. 528).

Significantly, the importance he accorded to the mechanisms of electrolysis to account for the discharge in gases enabled him to develop and work with the ideas of discrete matter and charge. The conflict between discreteness and continuum had to be resolved in the case of both matter and electricity. Maxwell's theory assumed a continuum displacement, not discrete charges. Thomson's theory of Faraday tubes helped him to think of charge as a phenomenon at their ends.

3.4   Opening the Cavendish to new researchers

In 1871, Cambridge saw the establishment of a hall of residence for women, which would eventually become Newnham Hall. Residents were entitled to attend lectures in the university but, *of course*, these women were not granted degrees or even formally admitted as registered members of the university. Little by little, and 'by the courtesy of the Examiners, women had been allowed to take the papers and these had been marked. Their place in the Tripos was not published in the University list, but it was not kept secret' (Thomson,

1936, p. 83). Embarrassment came when some of these women began to obtain high marks. The most impressive case happened in 1890, when Philipa Fawcett was ranked – *of course*, informally – above the senior wrangler, making national headlines in *The Times* (Warwick, 2003a, p. 218).

Lord Rayleigh was relatively forward-looking in regard to women's education and, from 1882, female students could attend demonstrations and lectures at the Cavendish. So, when Thomson was appointed director of the laboratory, women were a part, however small, of the landscape of the institution. One such young lady was Rose Paget, daughter of the Regius Professor of Physic, i.e., medicine, in Cambridge, who was interested in physics, although her formal training in advanced mathematics was quite limited. She attended some basic demonstrations at the Cavendish in 1887 and Thomson's own lectures in 1888. The letters between professor and student evolved from the very formal 'Dear Miss Paget' of March and August 1889 to the more friendly 'My dear Miss Paget' and to the unequivocal 'My darling Rose' the day before the formal engagement at the end of 1889 (Davis & Falconer, 1997, pp. xix – xx). The marriage took place on 2 January 1890, and the couple had two children: George Paget Thomson (G. P.), born on 3 May 1892 in the then family home of 6 Scroope Terrace, and in 1903, a daughter, Joan, in their new home on West Road.

Rose, familiarly known in Cambridge as 'Mrs J. J.', became an essential part of the laboratory, although not for her scientific work. She soon took on the task of helping J. J. to make the Cavendish an agreeable place where research was not incompatible with some degree of social engagement. She decided to contribute to the weekly discussions in the Cavendish with tea and biscuits she prepared, as well as making their family home as a sort of social annex the laboratory to entertain other fellows, professors, students, and visiting researchers. In his recollections, G. P. Thomson referred to weekly dinners, usually held on Saturdays, 'to which research students from the laboratory also came' (G. P. Thomson, 1966, p. 10).

The social dimension of the Thomsons' household became especially relevant in the years immediately after 1895, with the change in regulations at the University of Cambridge and their dramatic consequences for the Cavendish. German universities had been increasingly attracting graduate students from all over the world to not only improve their research skills but also to obtain Ph.D. degrees, which were very useful for obtaining university professorships. English universities were slow in implementing research degrees, thus discouraging young scientists in the British domains from improving their scientific training within the confines of the empire. The question turned into a political issue when it became clear that, as a consequence, British talent was contributing to the development of foreign (especially German) science. There was a clear need for change, especially in the Oxbridge system, to prevent this brain-drain.

In 1895, the Senate of the university decided that graduates from other universities were eligible to become research students in Cambridge, and new degrees were created: a Certificate of Research or a B.A., depending on the duration of residence at the university. Institutions like the Cavendish Laboratory, initially created for demonstrations to undergraduate students or for the personal research of the local professors, were the largest beneficiaries of the new regulations: they could now attract brilliant students from else-where, and thus enhance research with new and fresh ideas. Young men from England, Scotland and Ireland, as well as from Australia, New Zealand, Canada and the USA, immediately took advantage of this new possibility. The initial list includes people such as E. Rutherford, J. S. Townsend, J. A. McClelland, C. G. Barkla, J. A. Cunningham, and J. McLennan, many of whom eventually became major actors in the history of physics (Kim, 2002, p. 98).

Mr and Mrs J. J. were instrumental in making newcomers feel at home, since it was not always easy for them, without the long undergraduate training in the traditions of the colleges, to integrate into the insular world of Cambridge. Furthermore, not everyone was happy about having outsiders in their midst not only for fear that they might change the ethos of the university, but also because the skills and abilities of the newcomers were often superior to their own. Rutherford's letters sent to his fiancée in New Zealand on his arrival in Cambridge are a valuable contemporary snapshot of the Thomsons' care for the newcomers:

> I went to the lab and saw Thomson and had a good long talk with him. He is very pleasant in conversation and is not fossilised at all. As regards appearance he is a medium sized man, dark and quite youthful still; shaves very badly and wears his hair rather long. His face is rather long and thin; has a good head and has a couple of vertical furrows just above his nose. We discussed matters in general and research work and he seemed pleased with what I was going to do. He asked me up to lunch to Scroope Terrace where I saw his wife, a tall dark woman, rather sallow in complexion, but very talkative and affable. Stayed an hour or so after dinner … I like Mr & Mrs both very much. She tries to make me feel at home as much as possible and he will talk about all sorts of subjects and not shop at all (Rutherford, 8 December 1895, in Wilson, 1983, p. 64).

The new situation brought about an impressive flourishing of ideas and research topics. While, prior to 1895, Thomson and other fellows at the laboratory like

Glazebrook or Searle had undertaken their own private research and projects, the coming of new blood triggered the emergence of what some call the 'Cavendish school', a set of interrelated research topics to which J. J. contributed by giving a lot of advice and from which he, and everyone else at the Cavendish, also benefitted (Kim, 2002, p. 114–8; Rayleigh, 1942, p. 66). Actually, the new situation suited someone like Thomson: he was a vibrant source of ideas and suggestions which he would otherwise never have put into practice, partly for want of time, but also because of his lack of interest in finding precise proofs for and the final development of any given suggestion. As we shall see in the next section, J. J.'s long road towards a theory of conduction of electricity in gases reached its culmination at the end of the century partly thanks to the collaborative work in the 'Cavendish school', with the invaluable help of some of the new researchers, especially Rutherford (Falconer, 1985, p. 233).

At the same time, the laboratory had to accommodate increasing numbers of local undergraduate students attending lecture courses and demonstrations. By 1894, on average 200 students per term were filling the rooms of the Cavendish and space, which had always been limited, became a rare commodity. The university donated the land in Free School Lane adjacent to the existing building, and provided half the budget for a much-reduced construction plan. The other half, about £2000, came from the laboratory itself. This brings us to another aspect of J. J. as a science manager that has always been highlighted: his ability to run a demanding institution like the Cavendish, with increasing international prestige, on very limited economic resources, to the extent that some of his students and collaborators thought him extremely stingy. In fact, the university gave a very small annual allowance to the Cavendish, one that paid only for the salaries of the regular staff. The rest of the costs, including that of research materials, were met by the fees that students had to pay for the lectures and demonstrations they attended, and by ad hoc donations that J. J. managed to obtain. From these, he saved, in ten years, enough money to pay for half of the much-needed extension (Kim, 2002, pp. 79–83).

The formal inauguration of the new buildings took place in March 1896, just in time to meet the demands for space of the graduate newcomers. New buildings and new people coincided with the finding of new types of radiation, which that very year started to populate every physics laboratory in the world, including the Cavendish.

3.5   The corpuscle: notes from a 'discovery'

The folk history of science stresses crucial experiments, moments of revelation, great achievements by geniuses, and the like. Real scientific life is far more nuanced and therefore more interesting. The 'discovery' of the electron is no exception and, as Arabatzis argued, the very 'notion of the "discovery of the electron" has become a problem demanding explanation, as opposed to an explanatory resource' (Arabatzis, 2006, p. 69), in spite of the oft-repeated 'fact' that gives us a name, a date and a place: J. J. Thomson, in 1897, at the Cavendish Laboratory in Cambridge. In this section and the next, we shall look at the infancy of the electron *chez* Thomson, with only essential references to other participants in the birth of what we know as the first elementary particle. This story-line does, however, actually start with one of those few episodes in science that do have a straightforward name-date-and-place tag attached to it: the discovery of Roentgen rays.

In December 1895, Wilhelm Roentgen, a Professor of Physics in Würzburg, Germany, showed evidence of a new kind of radiation that, among many other unexpected properties, could produce images of the bones in a hand and of a compass needle inside an iron box. The famous picture of his own hand, ring included, triggered an immediate reaction in much of the scientific world (Pais, 1986, pp. 35–42). Cambridge was no stranger to this general response and J. J. Thomson managed to capitalize on his long experience with tubes and vacuum pumps to immediately work on X-rays and, indirectly, on cathode rays, since X-rays were obtained from cathode-ray tubes. J. J. described the general excitement over X-rays in the following terms in a paper in the February 1896 issue of *Nature*:

> The discovery of Prof. Röntgen of the rays which bear his name has aroused an interest perhaps unparalleled in the history of physical science. Reports of experiments on these rays come daily from laboratories in almost every part of the civilised world. A large part of these relate to the methods of producing Röntgen photographs, and the application of the 'new photography' to medical and other purposes. A considerable amount of work has, however, been done on the physical properties of these rays (Thomson, 1896a, p. 391).

In that same paper, published only two months after the first announcement of X-rays, he could already refer to experiments performed at the Cavendish on different aspects of Roentgen rays by himself, McClelland, C. J. R (sic) Wilson and Erskine Murray. In his presidential address to the British Association for the Advancement of Science in 1909, J. J. vividly compared the discovery of X-rays to that of 'the discovery of gold in a sparsely populated country; it attracts workers who come in the first place for the gold, but who may find that the country has other products, other charms, perhaps even more valuable than gold itself'. One of the side effects of this gold rush was the attraction of many workers to that which had been his research topic, the passage of electricity through gases. And this was the shift that started Thomson on the path towards the corpuscle: 'The study of gases exposed to Röntgen rays has revealed in such gases the presence of particles charged with electricity' (Thomson, 1909a, p. 10).

Thomson had hitherto performed only one experiment on cathode rays, consisting of a one-off study of their velocity. Unlike Arthur Schuster, whose work focused on the nature and behaviour of cathode rays, Thomson had seldom paid attention to this phenomenon. Despite having long investigated the nature of discharge in tubes, he encountered few manifestations of the presence of cathode rays, probably because he normally worked at higher pressures than Schuster (Falconer, 1985, p. 152). But, in 1894, he decided to analyse the speed of cathode rays in order to challenge experiments recently published by Hertz and his student Lenard, which seemed to show that cathode rays could pass through thin metal films and, *therefore*, that cathode rays were ethereal waves. Actually, as early as 1883, Hertz had studied the possible effects of electric fields on cathode rays, finding no deviation of the rays, which convinced him that the rays were some kind of undulatory phenomenon (Hertz, 1883). Had they been particles, they would have been deflected by the electric fields. Although these experiments were far from conclusive, they used to be cited by Continental physicists as an argument for the undulatory nature of cathode rays. As we shall see later, J. J. Thomson repeated these experiments only in 1897, proving that cathode rays were actually deflected and that Hertz's results were due to secondary effects.

But let us go back to 1894. In his studies on electrical conduction in gases, Thomson had paid much attention to the speed of discharge in tubes, a magnitude that was crucial to his modelling of a theoretical mechanism for discharge. That is why he was in a good position to undertake similar experiments with cathode rays, and to compare the new results with those previously gathered by other people. The velocity he obtained in 1894 was 'small compared with that with which the main discharge from the positive to the negative electrode travels between the electrodes', and he concluded that 'the velocity of the cathode-rays … agrees very nearly with the velocity which a negatively electrified atom of hydrogen would acquire under the influence of the potential fall which occurs at the cathode' (Thomson, 1894c, p. 364). Thus, his measurements on the speed of cathode rays were consistent with the traditional British understanding that they were corpuscular. For the wave explanation to be correct, cathode rays had to move at a speed close to that of light. This episode has often been quoted as evidence for Thomson's long engagement with a purported dispute between British and Continental physicists about the nature of cathode rays. But as Falconer convincingly showed, there was no such long-term warfare: Schuster was the only British physicist really interested in cathode rays before 1896, and other British physicists, Thomson included, paid scant attention to this phenomenon (Falconer, 1987).

Just like Thomson, Roentgen also tried to replicate Hertz and Lenard's experiments. He closely followed the latter's experimental set-up, except for one tiny detail. While Lenard had covered the cathode-ray tube with a thick coating so that cathode rays could only leave the tube through a tiny hole, Roentgen used only a thin sheet of black cardboard to stop the cathode rays. As it turned out, however, this shield was not enough to prevent some radiation from producing fluorescence on a photographic screen. In less than two months, Roentgen had studied the basic properties of this new radiation, or X-rays, presenting his results to the local Society for Physics and Medicine in Würzburg at the end of December, 1895. By the second week of January, 1896, the whole world had seen the spectacular pictures produced with Roentgen's rays in newspapers and magazines, and a number of physicists had started to work on the new radiation.

Thus, 1895 pumped new blood, new ideas and new resources into J. J.'s research programme on the conduction of electricity through gases. By the end of 1896, he had managed to develop a new theory of gaseous conduction based on results from his work *on* and *with* X-rays and on suggestions and help provided by his new research students, especially McClelland and Rutherford, with whom he also co-authored research papers in 1896 (Thomson & McClelland, 1896; Thomson & Rutherford, 1896). Questions about the origin, nature and properties of X-rays were all open.

Thomson soon found that X-rays ionized gases that would otherwise be non-conductors, which in turn was evidence for the high energy (higher than ordinary light) of such rays (Thomson, 1896b). This was important for three reasons: first, because it probably triggered in him the idea that, just as ordinary light was emitted by the vibration of the molecules of an element, X-rays might equally be the result of vibrations of 'finer pieces' of the gas molecules (Falconer, 1987, p. 265); second, because X-rays seemed to cause a dissociation similar to the one he imagined for electrification in gases: 'The passage of these rays through a substance seems thus to be accompanied by a splitting up of its molecules, which enables electricity to pass through it by a process resembling that by which a current passes through an electrolyte' (Thomson, 1896b, p. 276); and third, because it provided a method to quantitatively measure the intensity of X-rays or the conductivity of a gas: the rate at which a charged plate emitted electricity after being radiated with X-rays was proportional to their intensity.

Since 'Röntgenised' gases showed unexpected conductivity (Thomson, 1896c, p. 704), it was only natural that J. J., assisted this time by Rutherford, would undertake similar experimental studies to the ones he had done for over a decade, this time using X-rays, in order to get new ideas for building a definitive theory of electric conduction in gases. The radical novelty was that Thomson and Rutherford managed to put forward a *quantitative* theory of conduction, as opposed to the only *qualitative* suggestions of previous theories. With this, they could calculate, for instance, the number of particles into which the molecules of a gas would dissociate or the rate at which a 'Röntgenised' gas would leak (i.e., would cease to be a conductor). Following Thomson's understanding of the passage of electricity through gases as involving electrified particles, and with the new quantitative methods, Rutherford and Thomson inferred that 'the charged particles in the gas exposed to the Röntgen rays are the centres of an aggregation of a considerable number of [neutral] molecules' (Thomson & Rutherford, 1896, p. 402). In other words, Thomson thought that conduction was due to the movement of aggregations, the size of which seemed to be much *larger* than the ordinary structure of the gas molecules.

The most-articulated picture of Thomson's views around the autumn of 1896 can be found in a series of lectures he delivered in September at the University of Princeton, to which he was invited in the context of its 150th anniversary. It was his first trip to America, where he travelled in the company of his wife, and visited Johns Hopkins University, in Baltimore, before going to Princeton, the American university 'most reminiscent of Cambridge' (Thomson, 1936, p. 171). These lectures were published as a book in 1898 but, by that time, a lot had changed and Thomson had thoroughly revised the text. In the original manuscript, we find that J. J. Thomson greatly believed in his recent steps towards an explanation of electricity in gases through molecular aggregations, even if this meant a break with common explanations of electrolysis (where the carriers of electricity were obviously ions of atomic size). But, in hindsight, one can also feel huge tensions of physics in the making: we find constant references to experiments under way in the Cavendish and elsewhere, and the lack of a total and coherent theory of conductivity. As we have seen in the previous two paragraphs, he had reasons to think of structural units *smaller* than atoms to account for the origin of Roentgen rays, and for molecular aggregates *larger* than normal atoms to explain conduction in gases.

His next step, and one that would have unexpected consequences, was to see if this new theory of electric currents in gases was also valid, as he had hoped, for cathode rays, thinking that they might also be aggregates of molecules around a charged atom and, therefore, bigger than atoms. This hypothesis clashed, of course, with Hertz and Lenard's results, which he had tried to undermine two years earlier. And that is how, at the end of 1896, he started work on a topic to which, so far, he had paid scant attention. He also felt that the need to better understand the origins and nature of Roentgen rays made clarification on the nature of cathode rays urgent. Thus, contrary to traditional accounts of Thomson's supposed long interest on cathode rays, 'the discovery of X-rays did not divert attention from the cathode ray controversy, it virtually created it' (Falconer, 1987, p. 249; see also Turpin, 1980). J. J. Thomson, however, is partly responsible for the traditional history: his experiments on cathode rays were presented at a *public* event, the Friday Meetings of the Royal Institution, and J. J. justified his interest in the topic based on a supposed controversy between Continental and British accounts of cathode rays, with X-rays only as the catalyst for a renewed interest in them.

Following similar experiments by Jean Perrin some months earlier, Thomson's first goal was to verify that the negative electricity was actually carried by the cathode rays and not some secondary effect. This was important if he was to apply his dissociation theory of conduction to cathode rays. He soon also proved that the rays experienced magnetic deflection and that this was independent of the medium they traversed, which further supported their corpuscular nature. These two results, which were not particularly revealing or exciting, were presented to the Cambridge Philosophical Society in early February 1897 (Thomson, 1897a). The real change came in his Royal Institution Friday Meeting of 30 April, where he presented his first measurement of the charge-to-mass ratio for the carriers of cathode rays and speculated, for the first time, on the existence of corpuscles of very small mass. Most of that paper is a long and detailed explanation of his and Lenard's experiments on cathode rays proving their magnetic deflection and their identity as carriers of negative electricity. Hertz's old contention that cathode rays could not be deflected by electric fields and, therefore, could not be particles, was still rejected on the basis of arguments, not experiments; and Lenard's theory that cathode rays could traverse matter in ways that no known atom or molecule could still awaited explanation. The latter, together with the information from his own experiments, triggered the question he addressed in the last pages of the paper: 'These numbers raise a question ... and this is the size of the carriers of the electric charge. Are they or are they not the dimensions of ordinary matter?' (Thomson, 1897b, p. 12). To which he famously gave the following answer:

> Thus, from Lenard's experiments on the absorption of the rays
> outside the tube, it follows on the hypothesis that the cathode rays
> are charged particles moving with high velocities, that the size
> of the carriers must be small compared with the dimensions of

ordinary atoms or molecules. The assumption of a state of matter
more finely subdivided than the atom of an element is a somewhat
startling one; but a hypothesis that would involve somewhat similar
consequences – viz. that the so-called elements are compounds
of some primordial element – has been put forward from time to
time by various chemists. Thus, Prout believed that the atoms of all
the elements were built up of atoms of hydrogen, and Mr. Norman
Lockyer has advanced weighty arguments, founded on spectroscopic
consideration, in favour of the composite nature of the elements
(Thomson, 1897b, p. 13).

The reference to the *chemists* Prout and Lockyer should not be overlooked. The
possibility of some structure inside the atom accounting for chemical and electrical phenomena was not new to Thomson. He had toyed with the idea in his
vortex ring theory and again in the theories of electric discharge. His reference
to chemistry and spectroscopy, both part of his grand notion of the *Physical
Sciences*, can be read as his need for external support for his suggestion. But it
can also be interpreted as a typical attitude in J. J., who was always thinking
in terms of big pictures. As it often happened with him, his suggestion of a
tiny, subatomic corpuscle was immediately integrated into a major framework.
His tendency towards overarching theories that explained more than just one
phenomenon led him, on this occasion, to immediately extend his corpuscle
hypothesis into other areas.

At the same time, one should not be overawed, magnifying J. J.'s 1897 suggestion into some sort of prophetic insight. We have already seen him put
forward a number of loosely defined models and visual images with partial
explanatory power. At this point, the jump from an explanation of cathode rays
to an all-encompassing theory of matter seems to follow the same pattern. If
we look at his wording, his supposition was that 'the atoms of the elements are
aggregations of very small particles, all similar to each other; we shall call such
particles corpuscles, so that the atoms of the ordinary elements are made up of
corpuscles and holes, the holes being predominant' (Thomson, 1897b, p. 13). A
theory of matter based on corpuscles would include, as well, these *predominant
holes* about which nothing else was said.

Finally, his first estimation of the charge-to-mass ratio, which in turn would
give an indirect estimate of the mass of the corpuscles, was based on two measurements. First, from the experiments on magnetic deflection, a relationship
between the mass, the charge and the speed of the particles could be determined; and second, measuring the increase in heat of a given material under
the impact of cathode rays, he also obtained a relationship between the three

magnitudes. Cancelling out the speed of the rays, he obtained a value of the mass-to-charge ratio of 1.6×10^{-7}:

> This is very small compared with the value 10^{-4} for the ratio of the mass of an atom of hydrogen to the charge carried by it. If the result stood by itself we might think that it was probable that e was greater than the atomic charge of atom rather than that m was less than the mass of the hydrogen atom. Taken, however, in conjunction with Lenard's results for the absorption of the cathode rays, these numbers seem to favour the hypothesis that the carriers of the charges are smaller than the atoms of hydrogen (Thomson, 1897b, p. 14).

The third, and most frequently quoted paper of J. J. Thomson in 1897 was written in August and published in the October issue of the *Philosophical Magazine*. By now, he had deconstructed Hertz's experiment and proven that cathode rays were actually deflected by electric fields, final proof of their corpuscular nature. This, in turn, provided him with an alternative method for estimating the mass-to-charge ratio, obtaining a value of the same order of magnitude as the one previously obtained. This emboldened him to speculate not only on the nature of cathode rays but also on their role as universal constituents of matter, as well as suggesting his first tentative model of matter based on corpuscles. In the face of all the facts, he again used the authority of chemists and spectroscopists (basically Prout and Lockyer) to support his thesis that 'the atoms of the different chemical elements are different aggregations of atoms of the same kind', and he stated, with more emphasis than in April, that 'in the very intense electric field in the neighbourhood of the cathode the molecules of the gas are dissociated and split up, not into the ordinary chemical atoms, but into these primordial atoms, which we shall for brevity call corpuscles' (Thomson, 1897c, p. 311).

The sentence highlights the fact that J. J.'s theory of corpuscles was continuation of his previous theories of electric discharge. Corpuscles were obtained as the outcome in the dissociation process that, according to his theory, took place in gaseous electric conduction, but in a way he had so far not foreseen: that all gases would decompose into the same fundamental units and not into chemically distinct ions:

> Thus on this view we have in the cathode rays matter in a new state, a state in which the subdivision of matter is carried very much further than in the ordinary gaseous state: a state in which all matter – that is, matter derived from different sources such as hydrogen, oxygen, &c. – is of one and the same kind; this matter

being the substance from which all the chemical elements are built up (Thomson, 1897c, p. 312).

To avoid anachronism, we should not yet regard this 1897 corpuscle as the *first* elementary particle within a longer list to come. J. J. saw corpuscles as a new state of matter, as yet undefined, obtainable from all atoms and molecules and, therefore, being a subdivision of the chemical atom rather than a component of it. Furthermore, both the nature of matter and of charge and their mutual relations were ill defined. One possibility he briefly discussed was that the charge-to-mass relationship might be of the kind he considered in his 1881 paper, and later in his 1893 book, i.e., that the mass of the carriers of electricity in cathode rays was the 'quasi mass which a charged body possesses in virtue of the electric field set up in its neighbourhood' (Thomson, 1897c, p. 310). Furthermore, as we shall later see, the presence of corpuscles was not incompatible with, or a challenge to, his theory of Faraday tubes in the ether.

If in April he had speculated on atoms made out of corpuscles and 'holes', in this later paper, he put forward a more sophisticated idea of the atom, one in which atoms would be composed of these 'primordial atoms' arranged in electromagnetically stable configurations. He did not have a law for the forces between corpuscles, noting that there could be many theoretical models for such forces; but, true to his speculative tendency, he imagined that corpuscles in the atom might have similar arrangements to the stable configurations displayed by floating magnets: 'in this model the magnets arrange themselves in equilibrium under their mutual repulsions and a central attraction caused by the pole of a large magnet placed above the floating magnets' (Thomson, 1897c, p. 313). Like in 1881, when he studied the stability of certain configurations of vortex rings, these magnets also showed concentric stable arrangements of up to five elements, which 'seems to me to be suggestive in relation to the periodic law'.

3.6   Corpuscles and electrons

J. J. Thomson received the Nobel Prize in 1906 not for discovering the electron but for his long work on the conduction of electricity in gases. For some, this is an unfair decision: if he discovered the electron, why should he not be given the highest scientific honour for his longest-lasting contribution to science, i.e., for his discovery of the electron? The easy way to answer such a question is cite the problem the Nobel commission had in establishing the paternity of the electron, since Thomson was only one of many parents, including, at least, Pieter Zeeman, Walter Kaufmann, and Emil Wiechert. The commission *had* to give Thomson the prize, but could not take sides in priority disputes: that is why – the story goes – he was nominally awarded the Prize for *something else*. This is a very anachronistic interpretation, since it emphasizes what is important now for us and not what was important for the participants themselves at that time (1897 and 1906). As a matter of fact, in 1897, J. J. Thomson himself did not give priority to his supposed discovery of the first elementary particle, but continued working on a theory of electric discharge. At that time, the corpuscle hypothesis was, rather than a new theory of matter, a new and very important step towards a full theory of electric discharge, as his publications in the period 1898–1900 show. Only after he had finally solved the puzzle of electric discharge, around 1900, did he turn his attention to the corpuscle as the basic unit of matter. The corpuscle thus entered Thomson's world as the first (and, at the time, only) elementary particle only with the new century.

By that time, he had also obtained independent evidence of the charge-to-mass ratio of the corpuscles and for the inevitable conclusion that its value was necessarily due to the small mass of the corpuscles (and not to a possible larger charge of the negative ions). In fact, there was more speculation than heroic histories usually allow for in his 1897 corpuscles. To start with, he did not have convincing experimental evidence that the high charge-to-mass ratio of cathode rays was necessarily due to the smallness of the mass of the particles. Nor did he have evidence to suggest that corpuscles were universal constituents of matter. Both are, yet again, examples of Thomson's strong tendency to speculate based only on inconclusive results and a great deal of imagination. Only in the years between 1897 and 1900 were enough independent experimental results gathered to ground and confirm Thomson's corpuscle hypothesis.

Two papers from 1899 and 1900, and his 1903 book *Conduction of Electricity through Gases*, are the culmination of his theorising about electric conduction in gases. In these few years, J. J. Thomson relied heavily on the work of many of his research students, and, contrary to his previous practice, he now mentioned them explicitly in his papers. Thus, in his 1899 paper 'On the theory of the conduction of electricity through gases by charged ions', J. J. directly referred to the work of Rutherford and John Zeleny on the conductivity of gases under the radiation of Roentgen rays, to Rutherford's similar experiments on gases irradiated by uranium rays, to the changes of conductivity in a gas under the influence of a flame by McClelland and H. A. Wilson or under the influence of incandescent metals or of an arc discharge also by McClelland. All these investigations gave important information on the speed at which ions in a conducting gas moved, thus reinforcing, among other things, the asymmetry of such speeds for positive and negative electrification.

Contrary to the theory that had prevailed in the years prior to 1897, a discharge mechanism based on the movement and dissociation of ions could now explain such asymmetry, considering corpuscles as 'one of the small ions which are found in the cathode rays, and which we have reason to believe play an important part in all cases of electric discharge' (Thomson, 1900, p. 281). The much simpler mechanism he now put forward was

> that the ionization in the ordinary cases of discharge through gases is produced by the motion through the gas of *ions or corpuscles* already present in the gas, these ions or corpuscles under the action of the electric field acquire velocity and kinetic energy; and when this velocity or energy reaches a definite value which need not be the same for the positive and negative ion, these ions or corpuscles are able, by their collision with the surrounding molecules, to produce other *ions or corpuscles*. This dissociation may be directly due to the collision, or indirectly to rays like Röntgen rays produced by the collision (Thomson, 1900, pp. 279–80, my emphasis).

The free paths of the ions (positive) and corpuscles (negative) would be different due to their different masses, thus explaining the asymmetry in the discharge that had troubled him and many other researchers for such a long time. The corpuscle was finally the basic tool for the long-awaited mechanism of the transfer of electricity.

That electric discharge in gases was the apple of Thomson's eye is also clear in his 1903 book *Conduction of Electricity through Gases*, an encyclopaedic compendium of all knowledge available on the field, interpreted now in terms of the corpuscle theory. The book saw a second, largely revised, edition in 1906, and a third two-volume edition in 1928 and 1933. As with *Recent Researches*, *Conduction of Electricity* emphasized that 'the study of the electrical properties of gases seems to offer the most promising field for investigating the Nature of Electricity and the Constitution of Matter' (Thomson, 1903a, prologue to the first edition), a promising field he proudly located in the Cavendish, in the lectures and research he and his assistants were conducting. The whole book had two basic goals: one was to 'develope [sic] the view that the conduction of electricity through gases is due to the presence in the gas of small particles charged with electricity, called ions, which under the influence of electric forces move from one part of the gas to another', and the second to show that there was a strong asymmetry between positive and negative electrification and that this was fully explained by the existence of the negatively charged corpuscles. Corpuscles became, thus, the key to all electric phenomena, in a theory that, in his words, 'in many respects closely resembles that of the old "One Fluid Theory of Electricity":

> The 'electric fluid' corresponds to an assemblage of corpuscles, negative electrification consisting of a collection of these corpuscles: the transference of electrification from place to place being a movement of corpuscles from the place where there is a gain of positive electrification to the place where there is a gain of negative. Thus a positively electrified body is one which has been deprived of some corpuscles. These corpuscles may either remain free or get attached to molecules of matter with which they come in contact; thus positive electrification is always associated with ordinary matter, while negative electrification may or may not be, according as the corpuscles are or are not attached to molecules of ordinary matter (Thomson, 1903a, p. 63).

Note here a theme that will occupy us in the next chapter and the last: the distinction between 'ordinary' matter and the mass of corpuscles brings us to the two interconnected questions, 'What is electricity?' and 'What is matter?' Following the Maxwellian tradition, electric charge was a boundary phenomenon between ether and matter. Corpuscles were the carriers of negative electricity and the agents (by their excess or their deficiency) of all electrification in ordinary matter. But the way corpuscles were themselves electrified was, at this early stage, unclear. The hypothesis of the mass of corpuscles coming from their own electrification, which he had suggested in the 1897 *Philosophical Magazine* paper, and the relationship of such mass with the arrangement of ethereal Faraday tubes were still to be developed further.

In all his accounts, at least until 1911, Thomson used the word 'corpuscle' to refer to his massive carrier of negative electricity. 'Electron' had been used among physicists as a synonym of 'elementary quantity of electricity' since 1891, when George Johnstone Stoney had suggested the term at the Belfast meeting of the British Association for the Advancement of Science. The term was new, but not the idea: phenomena like electrolysis had long given evidence of a quantization of electricity, and Wilhelm Weber, perhaps the most influential authority on electricity the Continent, had toyed with the idea of small particles of unit electricity (Arabatzis, 2006, pp. 70–4). Towards the end of the century, the term had acquired in Cambridge a very particular meaning: in his search to formulate a final theory of electricity and matter, Joseph Larmor, whom we met beating Thomson as first wrangler, and was a lecturer in mathematics in Cambridge, had appropriated the term to denote mathematical singularities in the ether (Warwick, 2003b). Like Thomson and many other Maxwellians,

Larmor longed for a theory that explained the relationship between ether and matter; unlike Thomson, however, Larmor's career kept him away from laboratories and he did not realize the importance of chemistry in the pursuit of such a theory. Whereas Thomson's corpuscles were basically material units that carried the elementary charge of electricity, Larmor's singularities (or monads) in the ether *were* the actual electric charge. In the former, the main characteristic was the mass of the particle and the charge was left unexplained, while it was the other way around in the latter. In a way, one could say that Thomson was more faithful to Maxwell's legacy than Larmor, by sticking to a continuous electromagnetic ether without singularities. In other words, J. J.'s corpuscles had a mass and a charge that were, in unexplained ways, epiphenomena of the ether; while Larmor's electrons had introduced a new entity different from the ether or, perhaps more clearly, a new kind of *structured* ether.

Larmor's electrons were, like in the similar theoretical construction of the Dutchman H. A. Lorentz, inevitably identified with the smallest charged particle available at the moment, the ion of hydrogen (Darrigol, 1994). The first and very influential identification between Larmor's monads and J. J.'s corpuscles came in a paper written by George FitzGerald in the same issue of the *Electrician* where Thomson's April 1897 Evening Meeting was published. With this, FitzGerald's became the first authorized commentary of J. J.'s early hypothesis, one with which Thomson certainly did not agree. One of its most important characteristics, albeit still rather speculative in April 1897, was that the corpuscles were universal components of all matter, which FitzGerald thought could set Thomson 'within measurable distance of the dreams of the alchemists, and are in presence of a method of transmuting one substance into another' (FitzGerald, 1897, p. 104). But J. J. had not yet discussed in what way the corpuscles were actually the building blocks of matter, a question that was intimately related to the relationship between corpuscles, their electrification, and the ether.

4

On creeds and policies: the corpuscular theory of matter

4.1   What is an atom like?

Atoms; or is it molecules? The semantic fields of these terms were rather confused in the nineteenth century. In many instances, these two words were used almost synonymously but, more often than not, there was a lack of consistency in their usage, not only when comparing different research traditions, but also in their use by individuals. We have already found J. J. Thomson, for instance, in his discussion of the vortex ring theory, using the two terms equivocally. There are many historical reasons for this confusion, but two are particularly relevant here. First, in the nineteenth century, there were many theories of matter evolving in different intellectual arenas, especially one coming from the Newtonian and Laplacean tradition of physics, and another from the new chemistry of Lavoisier, two traditions that were miles apart in methods, conceptual formulation, and philosophical background. Second, the term 'atom' had a longer philosophical history than 'molecule', having acquired a sort of inertia that made it more difficult for the former to lose its malleability and acquire a more specific, well-defined 'scientific' meaning.

Maxwell gives us a good example of this confusion, with his two oft-quoted popular papers on the subject: the first, a general address to the British Association in Bradford in 1873, and his 1875 article for the ninth edition of the *Encyclopaedia Britannica*. The two papers are very similar, but the first is entitled 'Molecules' and the second 'Atom'. And his attempts to clarify the terminology are not always totally clear. Thus, for instance, he explains that by molecule he means the minimal unit of a particular chemical substance, either simple or compound, while atom 'is a body which cannot be cut in two', and 'if there is such a thing, must be a molecule of an elementary substance'

(Maxwell, 1874, p. 363). The two texts were meant for audiences not necessarily convinced about the reality of atoms, either because an indivisible unit of matter was a priori impossible, or because such small entities were not observable. With this in mind, he organised the papers so as to argue for the validity of atoms and/or molecules as hypotheses useful in many areas of physics and chemistry.

Maxwell's timeline of the modern atomic theory in those papers starts with the usual initial event: Dalton's revival and redefinition of the atomistic constitution of matter that had somehow been one of the fundamental tenets of early modern and Newtonian natural philosophy. Famously, Newton thought that 'It seems probable ..., that God in the Beginning form'd matter in solid, massy, hard, impenetrable, moveable Particles, of such Sizes and in such Proportion to Space, as most conduced to the End for which he form'd them' (Newton, 1730, p. 400). But it was with Dalton's law of chemical proportions that such atoms became entities *indirectly* measurable in the laboratory. The sizes and properties of such atoms were characteristic of each chemical element and, naturally, irreducible to each other. There were as many (and only as many) kinds of atoms as chemical elements one could isolate in the laboratory. Later, the Scot Thomas Thomson, one of the most influential British natural philosophers at the beginning of the nineteenth century, was instrumental in spreading Dalton's atomistic ideas. He was also a strong supporter of William Prout's idea that the masses of all atoms were exact multiples of the mass of the hydrogen atom and that, as a result, all atoms should be somehow composed of aggregations of that simplest atom or *protyle*. This suggestion was influential in Britain in the first third of the century, but further measurements proved it untenable (Nye, 1996).

Different kinds of atoms with different properties could have involved, in principle, the possibility of imagining them with different shapes, sizes or structures, giving them more realistic representations beyond the purely symbolic notation that alchemists had long used to denote the properties of the elements. But that did not happen, at least not to any significant extent, since the very existence of atoms was the main point of contention: talk of 'atoms' was basically a useful tool, as much as talk of 'chemical equivalents' or 'affinities'. And, as a matter of fact, the initial impetus for the atomic theory à la Dalton weakened with the growth of positivistic philosophy. In the words of Mary Jo Nye, 'from roughly 1840 to 1900, often in the name of creating a less hypothetical, more 'positive' science, the language of 'atoms' often came under fire and the language of 'equivalents' was in vogue, both in Great Britain and on the Continent, especially in France' (Nye, 1996, p. 44). Another point worth mentioning is that Dalton's theory had started as a way of unifying the more physical Newtonian tradition and the more chemical approach of Lavoisier and other – mostly French – natural philosophers. By the middle of the century, however, the two communities were, generally speaking, moving apart from each other. In Chapter 2, emphasis was placed on J. J. Thomson's institutional attempt to prevent such increasing separation, an aspect that will surface again in the next chapter with his support for the appropriation of the electron by the chemists.

To further complicate the possible emergence of a consistent atomistic science, we should also think of the increasing importance of kinetic theory in the nineteenth century. The work of Rudolf Claussius, in Zürich, and others had proven that heat could be statistically explained in terms of atomic movements. This was a new development of the Newtonian project that tried to explain everything in terms of components and their interactions. Rather than looking for new ethers, new fluids, and hidden properties, the kinetic theory promised to consummate a certain physical reductionism. But, unlike the chemical atom, Claussius' atoms were all the same, in the style of Newton and Laplace's scheme of things. Here, Maxwell is also central. With his mathematical *savoir-faire*, he could further develop the principles on which Claussius had built his heat theory and set the foundations for an overarching molecular theory of matter. In the lecture 'Molecules', Maxwell summarized the principles on which the atomic theory of heat was founded: the assumption that all matter is atomic, and that 'the molecules of all bodies are in motion, even when the body itself appears to be at rest' (Maxwell, 1874, p. 364). From the motions and mutual interactions of these 'molecules', and with their macroscopic physical constraints, one could quantitatively deduce some of the properties of the given body like pressure, temperature, specific heat, diffusion, thermodynamic equilibrium of mixtures, etc.

The development of statistical mechanics in the last third of the century, however, resulted in a further separation between the atoms of the chemists and of the physicists. The latter were imaginary, mathematical points with only mechanical properties. In Maxwell's words, there is 'no assumption with respect to the nature of the small parts – whether they are all of one magnitude. We do not even assume them to have extension and figure' (Maxwell, 1875, p. 451). His atoms were indistinguishable, as opposed to the ones used by the chemists: 'A molecule is that minute portion of a substance which moves about as a whole, so that its parts, if it has any, do not part company during the motion of agitation of the gas. The result of the kinetic theory, therefore, is to give us information about the relative masses of molecules considered as moving bodies' (p. 456). Atoms of hydrogen and of gold, for instance, were, for statistical purposes, basically the same. And, as a consequence, the atom of the kinetic theory did not ask for an internal structure of any kind. It was actually the ultimate unit from which to obtain other properties rather than an entity with properties.

Anti-atomism also arose from the problems that rigid atoms posed in explaining things that needed an internal structure for the atoms or, at least, diverse kinds of atoms: 'With such a structured atom, one could better understand the energy relations of chemical reactions; the indivisibility of Dalton's atoms severely strained Berthelot's credulity' (Nye, 1972, p. 9). Defendants of chemical atomism thus had to devise models of molecular structure in order to account, for instance, for the bonding and the behaviour of chemical substances. This became particularly important with the increasing development of organic chemistry and, especially, with the problems of isomerism. Thus, the atom of the chemist and the atom of the physicist seemed to follow divergent paths, which was, in turn, a challenge to any realist interpretation. Going back to Maxwell's lectures, 'If we suppose that the molecules known to us are built up each of some moderate number of atoms, these atoms being all of them exactly alike, then we may attribute the limited number of molecular species to the limited number of ways in which the primitive atoms may be combined so as to form a permanent system' (Maxwell, 1875, p. 480). But, he concluded, so far, nobody seemed to have a way to picture that.

There was, in fact, one such visual technique, although it was not universally accepted. There was a trend among British chemists to construct simple gadgets, for demonstration purposes only, to symbolize the molecular structure. Sets of balls with different colours representing different types of atoms, and sticks with which to attach these balls to act as the bond between them, were used in the classroom and in amateur lectures, and were even commercialized as toys for children. Of course, such devices were ridiculed by anti-atomists and regarded with deep reservation by many atomists, but this did not prevent their inclusion in some lecture rooms. They were particularly popular during the 1870s in Owens College, where J. J. received his first scientific training. There, Frankland and Schorlemmer used them regularly as a way to introduce students to the complexities of organic chemistry (Meinel, 2009). Certainly, there is no necessary connection, at least not one that can be documented, between these early molecular representations and J. J.'s later models of the atom, but it helps us to understand the broad British tradition of reasoning through images, a tradition in which Thomson received his early education and of which he became a paradigmatic instance.

Maxwell's lectures introduce us to another element that was beginning to provide information about the internal properties of atoms/molecules: spectroscopy. 'The molecule, though indestructible, is not a hard rigid body, but is capable of internal movements, and when these are excited, it emits rays, the wave-length of which is a measure of the time of vibration in the molecule' (Maxwell, 1874, p. 374). Each type of molecule had its own signature in the form of emitted radiation, irrespective of the history of the specific molecule. And that was, according to him, what set the nature of atoms outside the scope of evolution and, to a certain degree, even outside the scope of science: 'No theory of evolution can be formed to account for the similarity of molecules, for evolution necessarily implies continuous change, and the molecule is incapable of growth or decay, of generation or destruction', from which he inferred the manufactured nature of all molecules. And if molecules were manufactured through processes other than those 'we call natural', there was not much science could do to explain their origins. 'Thus we have led, along a strictly scientific path, very near to the point at which Science must stop' (p. 376). Beyond the obvious theological implications of this line of thought, there was an epistemological thread that prevented him from trying to imagine the internal structure of the atoms that form molecules: it is the molecules, and not atoms in isolation, which perhaps do not even exist, that we can study and imagine.

Not everyone thought this put an end to speculation about a possible internal structure of atoms. Also in 1873, the polemic Norman Lockyer brought forward his hypothesis of what he called 'celestial dissociation' to explain certain regularities in the solar spectrum (Lockyer, 1874, p. 493). His observations led him to assert that the spectra of chemical elements were dependent on temperature and that cosmic very high temperatures might dissociate atoms into supposed components in the same way laboratory high temperatures dissociated compound substances into their simple elements. As happened with many of his earlier hypothesis, this new theory of Lockyer's was also surrounded by controversy. Having received no professional training in science, and being well known for his often inaccurate chemical work in the laboratory, Lockyer's results were easily refuted and his general theories consequently dismissed. His dissociation hypothesis, dismissed by the Cambridge Professors of Chemistry, Liveing and Dewar, was no exception.

From this point of view, it is difficult to understand why Thomson chose to mention Lockyer's theory as support for his 'corpuscle' hypothesis in 1897. It might well be that, used as he was to building models without much ontological commitment to them, any support available for his corpuscle was welcome. Or we could think that, since the 1897 corpuscle was basically an instrument to explain the conduction of electricity in gases, and only indirectly a universal component of matter, the association he drew with the much-criticized theory of Lockyer reflects his being not yet totally committed to the latter aspect of the corpuscle. In any case, and with the background provided in this section, we can now try to explore J. J.'s path to the formulation of a theory of matter based on his corpuscles.

4.2   A world of electrons

Following John Heilbron, it is often repeated that 'the theory of atomic structure came into being with the discovery of the electron in 1897' (Heilbron, 1981, p. 14). That is only partly true, since J. J. Thomson's corpuscle was, in the first instance, a tool to explain the behaviour of cathode rays, and only later became the agent in a general theory of conduction of electricity. Only indirectly did it turn into the building block of matter: his corpuscular theory of electricity included a subatomic division into material carriers of electricity and, therefore, a theory of atomic composition. Once electricity was explained in terms of corpuscles, J. J. could look into the internal composition of atoms. At this stage, it does not surprise us that, following his monistic philosophy of nature, once he had corpuscles as components of the atom, it was only in terms of them – and nothing else – that he tried to build a complete theory of matter. The fact that all lines in the spectra of elements, and not only a few, showed the Zeeman effect was, for him, proof that corpuscles were present in large numbers in the atom.

Evidence for corpuscles came from several independent phenomena, and, by 1899, J. J. could legitimately claim that corpuscles were universal components of matter rather than an explanation of just cathode rays. Furthermore, he obtained an absolute value of the charge of corpuscles, not only of the mass-to-charge ratio, studying the electrification of metals under ultraviolet light. This gave more direct proof that corpuscles had the same electrification as the ion of hydrogen and, thus, that their mass was, with no little certainty, three orders of magnitude smaller than the smallest atom: 'From what we have seen, this negative ion must be a quantity of fundamental importance in any theory of electrical action; indeed, it seems not improbable that it is the fundamental quantity in terms of which all electrical processes can be expressed' (Thomson, 1899b, p. 565). The line of thought in this 1899 paper clearly shows that his theory of matter was an auxiliary hypothesis to his theory of electric conduction:

> These considerations have led me to take as a working hypothesis the following method of regarding the electrification of a gas, or indeed of matter in any state. I regard the atom as containing a large number of smaller bodies which I call corpuscles; these corpuscles are equal to each other … In the normal atom, this assemblage of corpuscles

> forms a system which is electrically neutral. Though the individual
> corpuscles behave like negative ions, yet when they are assembled in
> a neutral atom the negative effect is balanced by something which
> causes the space through which the corpuscles are spread to act as if
> it had a charge of positive electricity equal in amount to the sum of
> the negative charges of the corpuscles... On this view, electrification
> essentially involves the splitting of the atom, a part of the mass of
> the atom getting free and becoming detached from the original atom
> (Thomson, 1899b, p. 565).

This line of thought was apparent in his presentations of the corpuscle in the first years of the twentieth century (Thomson, 1901c). His main goal was to prove the existence of electrified material particles smaller than the smallest atom, and that these were the agents for the conduction of electricity in gases, metals, and radioactive substances, and even in the Aurora Borealis. His research, and that of many other researchers at the Cavendish, followed this line (Thomson, 1900, 1901a, b, 1902a, b, c). It looks as if having an atomic model was secondary to having a theory of electricity.

Once the existence of corpuscles was settled, J. J. began to explore all the possibilities of an entity that seemed to hold the key to the intimate connection between electricity and matter. It seemed to be the summit of his long-term project of understanding the relationship between matter and electricity – between matter and ether – that had been the driving force of his research programme on electric discharge in gases. It would also seem to support a monistic understanding of nature if one could not only explain atoms in terms of corpuscles but also understand corpuscles in terms of the ether. The monistic view of nature appeared to be only a step away.

The highlight of this period was his course at Yale, in May 1903, published immediately afterwards as *Electricity and Matter* (Thomson, 1904a). Again, here we find a story-line that seeks to demonstrate, first and foremost, the existence of corpuscles and their role in explaining electrification. As an illustration, Chapter 4, 'The atomic structure of electricity', comes before the chapter on 'The atomic structure of the atom'. Without the former, he could not argue for the latter. The contemporary reader may be further surprised by the content of the first three chapters, which are devoted to J. J.'s very own Faraday tubes. Corpuscles had not done away with them, but rather the contrary. Corpuscles were actually better explained in terms of Faraday tubes when supposing that the 'mass of a charged particle arises from the mass of ether bound by the Faraday tube associated with the charge' (p. 41). This was a generalization of his 1881 calculation of the apparent mass of a charged body due to the resistance in an electric field: then, he supposed the hydrodynamic resistance of a sphere in a fluid, while now he considered the mass of ether carried by the tubes per unit volume when moving at right angles to their axis.

A charged sphere at rest was postulated to be the origin of Faraday tubes extending uniformly in all directions. In this case, the electric mass would be zero, since – we should remember – Faraday tubes do have direction, and the effects of any tube are cancelled out by another tube in the opposite direction. When set in motion, the tubes would tend to have their axis perpendicular to the direction of motion of the sphere, thus creating much inertia. But since Faraday tubes repelled each other, there would be equilibrium between the two tendencies, giving us the measure of the bound mass carried by the tubes: 'When a Faraday tube is in the equatorial region it imprisons more of the ether than when it is near the poles, so that the displacement if the Faraday tubes from the pole to the equator will increase the amount of ether imprisoned by the tubes, and therefore the mass of the body' (Thomson, 1904a, p. 43). Thomson went on to show that the 'assumption that *the whole of the mass is due to the charge*' (p. 48, emphasis in the original), to which he was highly inclined, was more than likely.

If the mass of the moving charged sphere was due to the mass of the ether carried along by the Faraday tubes, this would mean that, in principle, the mass of any charged particle extended indefinitely with the tubes. That was not a problem, he argued, taking into account that in small particles like the corpuscles, the mass of ether carried by the tubes decreased according to the fourth power of the distance from the particle, and thus, 'all but the most insignificant fraction of mass is confined to a distance from the particle which is very small indeed compared with the dimensions ordinarily ascribed to atoms' (Thomson, 1904a, p. 50). And from this he advanced his dreamt-of-ontology: '… the *whole* mass of any body is just the mass of ether surrounding the body which is carried along by the Faraday tubes associated with the atoms of the body. In fact, that all mass is mass of the ether, all momentum, momentum of the ether, and all kinetic energy, kinetic energy of the ether' (p. 51).

And what was the relationship between these Faraday tubes and the charges of electricity? Only that the latter *were* 'the beginnings and the ends' of these tubes. Here language fails him, since he was actually saying that there was no clear distinction between mass, charge, and ether. If the mass of a particle expressed the mass of ether carried by Faraday tubes, electrification was the phenomenon at the extremes of tubes. 'If this view of the structure of electricity is correct, each extremity of a Faraday tube will be the place from which a constant fixed number of tubes start or at which they arrive' (Thomson, 1904a, p. 71).

As for the structure of the atom, J. J. was once again very cautious about the role of corpuscles: 'It may thus not be superfluous to consider the bearing of the existence of corpuscles on the problem of the constitution of the atom'. This was not simply a rhetorical device to caution his audience, but his actual line of thought. Faraday tubes were real, the ether was real, corpuscles were an explanation of electrification and, perhaps, of the structure of matter. This had been the order of his preferences in research and this was the order, in 1903, of his pedagogical presentation of corpuscles. Infact, at this stage, he did not have, strictly speaking, a model for the atom, but a research programme: 'although the model of the atom to which we are led by these considerations is very crude and imperfect, it may perhaps be of service by suggesting lines of investigation likely to furnish us with further information about the constitution of the atom' (Thomson, 1904a, p. 92).

And what was this atom like? J. J. Thomson thought of it as a collection of what he called doublets, 'with a negative corpuscle at one end and an equal positive charge at the other, the two ends being connected by lines of electric force which we suppose to have a material existence' (Thomson, 1904a, p. 93). Thus, the atom appeared as an assemblage of Faraday tubes with one very condensed end, forming the individual corpuscles, and another end spreading over a comparatively much larger space. In this way, he could imagine that 'the quantity of ether bound by the lines of force, the mass of which we regard as the mass of the system, will be very much greater near the corpuscle than elsewhere' (p. 94), or, in other words, that the mass of the atom could be considered as the sum of the masses of what we see as corpuscles. This gives us an atom about which we can speak at different levels. Deep down, it is basically an assemblage of Faraday tubes; but, at the next level, we can visualize it as an assemblage of corpuscles in a sea of positive electrification. With the latter image, he discussed the problem of the stability of such a system and the light this threw on chemical bonding and also on radioactivity.

Following the highly speculative tenor of this book, Thomson went on to suggest the concept of 'corpuscular temperature', a statistical measure of the movement of corpuscles inside the atom, analogous to the molecular temperature in the kinetic theory of gases. This temperature would account for the possibility of chemical components as well as for the stability and life of atoms. In principle, radiation emitted by moving corpuscles is a problem for the stability of atoms, but there was a possible way out of this problem: 'we must remember, too, that the corpuscles in any atom are receiving and absorbing radiation from other atoms' (Thomson, 1904a, p. 108). In a 1903 paper, 'The magnetic properties of systems of corpuscles describing circular orbits', he had dealt with the problem of the stability of moving electrons and shown that, in a rotating ring of corpuscles, the radiation emitted by the individual electrons interfered with that of the other electrons, thus diminishing their energy loss due to radiation (Thomson, 1903b). In his typically loose way, Thomson assumed that this phenomenon would, in a system with a large number of corpuscles, cancel out all radiation, thus making the atom a stable system.

Finally, he considered the geometry of the arrangement of corpuscles inside the atom. The simple cases of two, three, and up to seven or eight corpuscles, could be calculated, giving symmetrical arrangements on the surface of an inner sphere concentric to the sphere of the positive electrification. But for larger numbers, the corpuscles broke up into several groups. Thinking that the simplest atom, that of hydrogen, would have on the order of 10^3 corpuscles, 'the problem of finding the distribution when in equilibrium becomes too complex for calculation', and he suggested turning to the 'simple and beautiful' experiment of the arrangement of magnets floating in a vessel of water (Thomson, 1904a, p. 117). As Figure 4.1 shows, the arrangement of the magnets in stable layers is suggestive of the atomic properties according to the 'Periodic Law', by

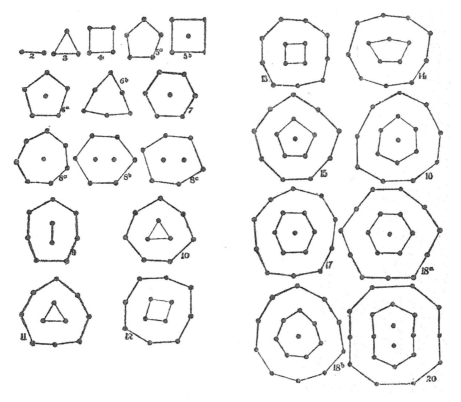

Figure 4.1. The arrangement of magnets on a fluid was suggestive of the properties of atoms in the Periodic Table (Thomson, 1904a, p. 115). J. J. Thomson took the idea of Mayer's arrangement from a loose model for radioactivity by Lord Kelvin (W. Thomson, 1902).

which a minor increase in the mass of the elements does not render them similar in chemical properties, but, rather, we have to wait for a full cycle to find the same properties. Not surprisingly, J. J.'s model of the atom was immediately relevant to chemical as well as physical phenomena.

Many historians overlook these lectures Yale and repeat the claim that the first time J. J. Thomson developed his atomic model was in a *Philosophical Magazine* paper of 1904 (Thomson, 1904b). Certainly, in that paper, as well as in a public lecture on 10 March 1905, J. J. described his atoms as an assemblage of corpuscles, without reference to their intrinsic nature as endpoints of Faraday tubes. His starting point was then the existence of corpuscles, the only building block from which he constructed his atomic model, without reference to their intimate nature: 'if the corpuscles form the bricks of the structure, we require mortar to keep them together. I shall suppose that positive electricity acts as the mortar, and that the corpuscles are kept together by the attraction of positive electricity' (Thomson, 1905b, p. 1). Faraday tubes and corpuscles were entities at different ontological levels and J. J. thought it would be more helpful to present his model of the atom on the basis of corpuscles, leaving their nature for other, more specialized, audiences. In this way, he managed to present his atom in a fashion that was very appealing to chemists as well as to physicists. J. J. wanted to be very clear from the outset that his corpuscular atom was the atom of both the physicists and the chemists and, thus, he could claim to have found the final link between the two scientific traditions.

4.3   Psychic research

Ordinary material systems must be connected with invisible systems which possess mass whenever the material systems contain electrical charges. If we regard all matter as satisfying this condition we are led to the conclusion that the invisible universe – the ether – is to a large extent the workshop of the material universe, and that the phenomena of nature as we see them are fabrics woven in the looms of this unseen universe (Thomson, 1907f, p. 21).

This is J. J. Thomson at the end of a public lecture in Manchester, in November 1907, in which he argued that the latest discoveries in electricity proved a deep metaphysical unity in the world, one that stemmed from the real existence of the ether and from the fact that all phenomena in Nature were a result of matter in movement. Two analogies permeate the rhetoric of this lecture: the machine-like fabric of the world, and the existence of an unseen universe that keeps that machine in productive movement. The first comparison is natural at the centre of British manufacturing industry. The latter resonates with that best-seller by Tait and Stewart that had argued for a strong link between the science of energy and the existence of spirits and their action in the visible universe.

This mention of an *unseen universe* is not an isolated occurrence. In his auto-biography, J. J. felt the need to write one full chapter, albeit the shortest, on psychic research, a topic to which he had devoted intellectual attention and political support, especially through his membership of the British Society of Psychical Research (SPR), of which he was even vice-president for some time. The society, formally founded in 1882, was intended as a scientific response to the multitude of groups and associations interested in all kinds of paranormal phenomena in Victorian Britain. Interest in spiritualism and related issues had all sorts of motivations: as a way to prove the reality of an afterlife and the need for religion, as a means to challenge the authority of the established Anglican Church, as a response to the threat of increasing materialism, and as a way to extend the scientific ethos to the matters of the mind (Gauld, 1968; Haynes, 1982; Oppenheim, 1985). The SPR was particularly cautious about the status of spiritual and psychic phenomena and its aim was 'to examine without prejudice or prepossession and in a scientific spirit those faculties of man, real or supposed, which appear to be inexplicable in terms of any generalized hypotheses' (quoted in Gauld (1968) p. 137). The society was well respected among Cantabrigian academics since amongst its most enthusiastic driving forces were two Cambridge dons: Henry Sidgwick and Frederick Myers.

Sidgwick was a Trinity graduate in Classics who became lecturer in moral philosophy and, eventually, Knightsbridge Professor of Philosophy in Cambridge. In 1869, he resigned from his appointment at Trinity College, since he no longer felt he could assent to the 39 articles of faith that fellows had to sign, although the College found extraordinary ways to keep him until 1882, when regulations were changed, and he was readmitted as an ordinary fellow. Ever since his undergraduate years, Sidgwick had been involved in psychic research through the Cambridge 'Ghost Society', and, as a mature philosopher and Anglican apostate, he saw in spiritualism a possible way to support Christian morality without assent to its theological content (Oppenheim, 1985, pp. 113–16). In J. J. Thomson's words, 'he was one of the most brilliant talkers of his time … [and] the most brilliant in Cambridge' (Thomson, 1936, p. 294), and greatly involved in reforms at the University. Particularly important was his and his wife's work in connection with women's education and the creation of Newnham College.

Sidgwick became the first president of the SPR. 'He was an ideal president for such a society, absolutely fair and unbiased and critical' (Thomson, 1936, p. 299): he was a highly respected and honest man who did not hide his many disappointments in his search for psychic and spiritual evidence. In fact, Sidgwick was 'notoriously unlucky as a psychical researcher', and while people like Crookes, Lodge, and Wallace saw indubitable evidence of some paranormal phenomenon, Sidgwick, 'in spite of repeated trials, … never witnessed anything' (Crookes, quoted in Oppenheim (1985) p. 124). His interest in the subject was challenged by many disappointments, turning him into a moderate agnostic and, thus, a respectable president of the society.

Myers was by far more enthusiastic than Sidgwick and the one who brought J. J. into the SPR: 'in the nineties, at the instance of F. H. W. Myers, I attended a considerable number of séances at which abnormal physical effects were supposed to be produced' (Thomson, 1936, p. 147). Although no longer a lecturer Trinity College, stayed in Cambridge as part of the intellectual elite, making himself a name as a poet, critic and essayist. He first became interested in psychic and paranormal research through the influence of Sidgwick, his undergraduate tutor at Trinity in the 1860s, but his interest surpassed that of his tutor's from the 1870s onwards. Through the work of the SPR, he became increasingly convinced of the importance of hypnotism in developing a science of the psyche, and he developed the concept of the 'subliminal self', 'the boldest and best known of the contributions that psychical research made to psychology before World War I' (Oppenheim, 1985, p. 254).

In his recollections, J. J. seems to have mixed feelings about the activities of the SPR. He was clearly disappointed by the fact that 'at all but two of those [séances] I attended nothing whatever happened, and in the two where something did there were very strong reasons for suspecting fraud' (Thomson, 1936, p. 147). In spite of that, J. J., as well as Lord Rayleigh and some other physicists in Cambridge, 'maintained a deep interest in the society's work but conducted only occasional investigations into psychical phenomena' (Noakes, 2005, p. 426): they were observers and not active actors in this research. Thus, they had no serious grounds to dismiss an activity to which they were, at least in principle, not opposed. Their interest was probably not so much in psychology, let alone a belief in spirits and ghosts, but in the possibility of extending the domains of physics to the study of the mind. J. J. had a more positive attitude towards telepathy, 'another branch of psychical research which may be connected with physics', and of which he had witnessed some positive instances. By the time he wrote his memoirs, J. J. still thought that, 'in my opinion the investigation of short-range thought transference is of the highest importance' (Thomson, 1936, p. 154).

One very popular story in Cambridge concerned the visit in the summer of 1895 of an illiterate Italian peasant, Eusapia Palladino, who had acquired a name as medium. Invited by the SPR after Lodge and Myers were convinced of her powers at a séance in France (also attended by the more sceptical Sidgwick), Palladino performed a number of what many saw as deceptive tricks. Apparently, her behaviour 'stimulated the prejudices latent in the Sidgwick group', J. J. included. In a most ironic paragraph, J. J. also describes the case of the famous Madame Blavatsky in her visit to Cambridge:

> One of my most interesting experiences was a séance when nothing at all happened... She said at the beginning that her Mahatma in Tibet would precipitate a message, a cushion and a bell, and we sat waiting for, I should think, more than an hour, and nothing whatever arrived. The medium was not in the least abashed. She took the offensive, said it was all our fault, that our scepticism had created an atmosphere impenetrable to anything spiritual. She was a short and stout woman with an amazingly strong personality, very able and an excellent speaker. So well did she speak that she convinced the great majority of the audience that the failure was their fault, and they went away thoroughly ashamed of themselves for having spoiled what would otherwise have been a most interesting experience (Thomson, 1936, pp. 153–4).

In spite of this account, J. J. still thought that good work was done by the SPR: 'This work has not been wasted. To put its claims at the very lowest it is surely a great thing to have created an organisation for collecting and testing these abnormal phenomena and thereby to go far to ensure that no genuine ones will escape discovery' (Thomson, 1936, p. 299). As Richard Noakes argued, Thomson's interest in and membership of the SPR problematizes the positivistic view that, by the turn of the century, there was a clear-cut definition of the limits of physics. The history of psychic research needs to be seen 'as an episode in late-classical physics' rather than as something alien to it (Noakes, 2008, p. 326).

4.4   The collapse of a dream

Between 1903 and 1906, J. J. Thomson published a number of papers on the physical and chemical properties of his atomic model. The arrangement of corpuscles in the atom was the crucial point in determining such properties, and many of his papers dealt with issues such as 'The magnetic properties of systems of corpuscles describing circular orbits' (Thomson, 1903b), 'On the vibrations of atoms containing 4, 5, 6, 7, and 8 corpuscles and on the effect of a magnetic field on such vibrations' (Thomson, 1905a), and 'A theory of widening of lines in the spectra' (Thomson, 1906a), in which he assumed that the widening of the spectral lines was 'due to the effect of resonance between systems which if free from each other's influence would vibrate in the same period' (p. 318). Thomson's excitement in these years was manifested, for instance, in a letter to Rutherford, in which he expressed his 'hopes of being able to work out a reasonable theory of chemical combination and many other chemical phenomena', including radioactivity, based on his atomic model (J. J. Thomson to Rutherford, 16 February 1904, CUL Add 7653, T23).

If the arrangement of corpuscles in the atom was going to be a factor relevant to the chemical properties of atoms, he had first to determine their approximate number. Every model showed that corpuscles were distributed in different layers and that it was only the outer ones that were really significant for chemical properties. Calculation of such distributions was totally impracticable if their number was of the order of the thousands, and, as we saw earlier, one had to rely on experimental models like the one provided by Mayer's magnets. That is why J. J. prioritized the problem of the number of corpuscles in the atom, eventually finding three independent ways to determine the figure. In the first, he assumed that optical dispersion in monatomic substances was the result of a polarization of the atom, from which he could gain information about the mass of both positive and negative electrification. The second came from a theory of his and C. G. Barkla, one of the first Cavendish researchers to systematically study Roentgen rays, by which the energy of scattered X-rays was proportional to the density of corpuscles in a gas. Measuring the dispersion in air of known density, Barkla calculated that there was only room for an average of 25 corpuscles per molecule of air. Finally, and on the hypothesis that beta absorption was a result of the collisions of radioactive particles (themselves also corpuscles) with the corpuscles in the absorbing substance, J. J. also reached the conclusion that the number of corpuscles in atoms had to be of the order of magnitude of the atomic weight.

Interestingly for the contemporary reader, the three methods relied on the assumption that the laws of optical dispersion, X-ray diffraction and beta absorption were the same outside and inside the atom, something that the then emerging quantum theory was only beginning to challenge. In any case, the three estimates coincided enough that 'the evidence at present available seems ... sufficient to establish the conclusion that the number of corpuscles is not greatly different from the atomic weight' (Thomson, 1906b, p. 769). That is how his monistic-corpuscular model came to an end, and a new set of problems arose. If most of the atomic mass was not due to corpuscles, it had to be related to the positive electrification. And no less importantly, with a small number of corpuscles, the stability of the atom was under threat. It was only the large number of corpuscles that had so far prevented the atom from collapsing due to the radiation of moving corpuscles.

This radical change of affairs came at the time J. J was giving a series of lectures at the Royal Institution in London. The course, published in 1907 as *The Corpuscular Theory of Matter*, is a good exposition of J. J.'s uses of the corpuscle at the time. As in previous presentations, his first goal was to convince his audience of the existence, properties, and nature of the corpuscles, and their usefulness in accounting for many physical and chemical phenomena. Famously, however, he reiterated the heuristic character of any model, including a model based on corpuscles, with the following:

> The theory of the constitution of matter which I propose to discuss in these lectures, is one which supposes that the various properties of matter may be regarded as arising from electrical effects. The basis of the theory is electricity, and its object is to construct a model atom, made up of specified arrangements of positive and negative electricity, which shall imitate as far as possible the properties of the real atom ... The theory is not an ultimate one; its object is physical rather than metaphysical. From the point of view of the physicist, *a theory of matter is a policy rather than a creed*; its object is to suggest, stimulate and direct experiment (Thomson, 1907a, p. 1, my emphasis).

His emphasis on policies, not creeds, is totally consistent with what we saw in the previous chapter. What should perhaps be emphasized is that the heuristic character of physical theories was also true for corpuscles and corpuscular models of matter. Corpuscles and their behaviour were not the last word on what matter was. As we have already seen, and shall see again in the last chapter of the book, explanations in terms of corpuscles were totally compatible with the existence of a more fundamental continuum in matter. Corpuscles were not an alternative mutually exclusive to ether and the continuum, but a different, more phenomenological, layer of explanation. In fact, he maintained that the optimal theory of matter was the one based on vortex rings, but that 'The simplicity of the assumptions of the vortex atom theory are, however, somewhat dearly purchased at the cost of the mathematical difficulties which are met with in its development; and for many purposes a theory whose consequences are easily followed is preferable to one which is more fundamental but also more unwieldy' (Thomson, 1907a, p. 2).

That corpuscles were one possible, but not the ultimate, explanation of matter was made clearer when accounting for the origin of their mass. We have already seen that J. J. considered their mass as arising from the movement of the ether structured in Faraday tubes. But now we find J. J. drawing a surprising conclusion, one that might sound prophetic of the emergence of de Broglie's principle 20 years later: 'hence from our point of view, each corpuscle may be said to extend throughout the whole universe, a result which is interesting in connection with the dogma that two bodies cannot occupy the same space' (Thomson, 1907a, p. 34). I do not mean here, of course, that J. J. Thomson somehow predicted de Broglie's principle, but only that, as we shall see in Chapter 6, when de Broglie's principle appeared on the stage, J. J. was ready and happy to accept it in terms of his own ether-based vision of matter.

The corpuscle had come as the answer in a long research project on the conduction of electricity in gases. But what about conduction in solids? It was 'somewhat remarkable' that 'the passage of electric conduction through metals … by far the most familiar case of electric conduction' was the less understood in terms of the 'mechanism by which conduction is effected' (Thomson, 1907b, p. 455). And to this topic J. J. devoted a third of his Royal Institution lectures, developing two alternative theories of electric conduction in metals. The first, essentially the same as that first developed by Paul Drude in 1900 (Drude, 1900a, b), assumed that the corpuscles from the atoms in the conductor material were free and formed a kind of gaseous sea of corpuscles in thermal equilibrium: 'these corpuscles can move freely between the atoms of the metal just as the molecules of air move freely about in the interstices of a porous body' (Thomson, 1907a, p. 51). In the presence of an external electric field, the negative carriers of electricity would move without affecting the atoms of the conductor. In this model, calculations were analogous to the ones in the kinetic theory of gases: the movement of corpuscles depended on the external electric field, but also on their mutual collisions, and from this one could calculate and relate the mean free path of a corpuscle, the specific conductivity, and the number of corpuscles per unit volume.

Comparison with experimental data gave reasonable corroboration of this first theory with phenomena like the Hall effect, the Peltier effect, and the Thomson effect (the last named after William Thomson, Lord Kelvin), but gave totally inconsistent results when relating the estimated number of corpuscles to the conductivity of metals in fast-changing electric fields. In order to account for this inconsistency, J. J. Thomson introduced a modification to the model. In what he called the 'modern theory of electrical conductivity of metals' (Thomson, 1907b), corpuscles were bound to their atoms, which could be represented as electric doublets. In the presence of an electric field, these doublets would tend to organise themselves along a line 'much in the same way as the Grotthus chains in the old theory of electrolysis'. By this theory, 'the corpuscles are supposed to be pulled out of the atoms of the metal by the action of the surrounding atoms' when these were in line (Thomson, 1907a, p. 86), and therefore there was no need for a large number of free corpuscles in movement in order to get electric conduction.

This new theory of electric conduction in metals is clearly reminiscent of his theories on electrolysis and on gas discharge. But there is another aspect worth highlighting. In all these theories, J. J. Thomson needed to deal, once again, with the asymmetry between positive and negative electrification. The latter resided in the corpuscles, and the former in 'molecules' with a deficit of corpuscles. But the demonstration that corpuscles were not sufficient to account for the mass of atoms and molecules made elucidation of the nature of positive electrification pressing. To this he shifted his attention, starting in 1907.

4.5   The carriers of positive electricity

From 1907 to 1910, J. J. Thomson did not speculate any further on his atomic model. He did not change or abandon it, but he certainly had no further arguments to give it more consistency, due to the problem with positive electrification and the challenge it posed to his monistic view. At the same time, and in the absence of anything more convincing, J. J. still held the basic tenets of the atomic model which guided his early research in positive electrification. This may also explain why the textbook history of science moves from Thomson's 'plum-pudding' model to Rutherford's 1911 nuclear atom without drawing attention to the huge problems of the former model that became apparent from 1907 onwards. Furthermore, as we shall see in the next chapter, J. J. addressed again the question of the structure of the atom after 1910 only as a tool to account for the nature and behaviour of light and radiation. Then, his various and contradictory speculations never went beyond the limited realm of auxiliary and ad hoc hypotheses.

His monistic theory of everything had been just one step away, as he intimated to Oliver Lodge in a 1904 letter:

> I have however always tried to keep the physical conception of
> the positive electricity in the background because I have always
> had hopes (not yet realized) of being able to do without positive
> electrification as a separate entity and to replace it by some property
> of the corpuscles. When one considers that all the positive electricity
> does on the corpuscular theory is to provide an attractive force to
> keep the corpuscles together, while all the observable properties of
> the atom are determined by the corpuscles, one feels, I think, that the
> positive electrification will ultimately prove superfluous and it will be
> possible to get the effects we now attribute to it from some property

of the corpuscle (J.J. Thomson to Lodge, 11 April 1904, quoted in
Rayleigh 1942, pp. 140–1).

Already in 1899, when determining the absolute value for the mass of the carriers of negative electricity, J. J. tried to determine the same value for the positive electrification. The asymmetry between the two was very obvious, and the
conclusion he drew was that 'the carriers of positive electricity at low pressures
seem to be ordinary molecules while the carriers of negative electricity are
very much smaller' (Thomson, 1899b, p. 557). With this conservative attitude,
Thomson hoped not only to keep the widely accepted idea that the positive ion
of hydrogen was the smallest positive particle, in contrast with his highly revolutionary hypothesis of the existence of the unprecedentedly small corpuscles,
but also to take a step in the direction of his monistic agenda.

Kanalstrahlen, as they had been known since their discovery in 1886, were
rays of positive electrification coming through a hole in the cathode of a
discharge tube. J. J. renamed them positive rays and, in his words, 'these
rays seem to be the most promising subjects for investigating the nature
of positive electricity' (Thomson, 1907c, p. 2). The experimental set-up he
used was reminiscent of his work on cathode rays: the pencil of positive rays
was subjected to electric and magnetic fields, because he hoped that their
deviation would give information on their nature and behaviour. Wilhelm
Wien had already studied the basic charge-to-mass ratio of positive rays at
the turn of the century. His results showed that positive rays had no definite
value of the charge-to-mass ratio except for an upper limit: that of the H^+ ion.
J. J. designed his experimental set-up accordingly, expecting a range of values
rather than a specific one. To this end, he set the electric and magnetic fields
parallel to each other, so that they would produce perpendicular deflections.
With this, he expected to observe a parabola on the fluorescent screen at the
end of the tube, with a different characteristic parabola for each possible
kind of positive ray.

His first observations, although very provisional, were not in accord with
his expectations. What he saw on the fluorescent screen was more like a
straight line with a sharp upper limit corresponding to the value of charge-
to-mass ratio for the positive ion of hydrogen, and not a parabola (Figure
4.2). He also observed a few negative dots on the screen with the same range
of values for the charge-to-mass ratio (and, thus, clearly not corpuscles). His
first and very hasty conclusion was that all the positive particles were ions of
hydrogen, only their charge was statistically not constant. Since the corpuscles were the only bearers of electrification, the varying values of the charge-
to-mass ratio for the positive particles should be due to a variable degree

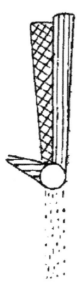

Figure 4.2. Under the influence of parallel magnetic (B) and electric (A) fields, the positive rays were expected to form a characteristic parabola on the screen of the kind $y^2 = (B^2/A)\,(e/m)\,x$, for each value of the charge-to-mass ratio but what he detected was not exactly a parabola (Thomson, 1907d, p. 568). Courtesy: Taylor and Francis, Ltd.

of electrification rather than a variation of their mass. Actually, Wien had, since 1898, found similar values for the charge-to-mass ratio of Canal rays (or Kanalstrahlen), but he had toyed with the idea that these had a fixed charge and variable mass (Wien, 1898). To support his hypothesis, J. J. made up a completely ad hoc mechanism that could explain the statistical variation in the charge of the H^+ particles:

> Suppose that some of the particles constituting the positive rays, after starting with a positive charge, get this charge neutralized by attracting to them a negatively electrified corpuscle, the mass of the corpuscle is so small in comparison with that of the particle constituting the positive ray that the addition of the particle will not appreciably diminish the velocity of the positive particle. Some of these neutralized particles may be positively ionized again by collision while others may get a negative charge by the adhesion to them of another corpuscle, and this process might be repeated during the course of the particle. Thus there would be among the rays some which were part of their course unelectrified, at other parts positively electrified, and at other points negatively electrified (Thomson, 1907c, p. 10).

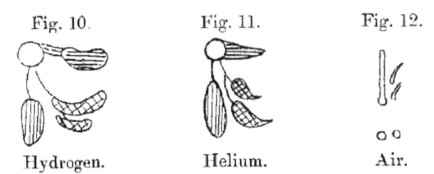

Figure 4.3. The fluorescence he obtained for different gases was different to what he expected. Only helium seemed to give patterns close to his expected parabolas (Thomson, 1907d, p. 573). Courtesy: Taylor and Francis, Ltd.

In other words, positive rays, in their passage through the electric and magnetic fields, would acquire and lose negative corpuscles present in the tube, thus causing them to experience almost random deflections, except for a maximum corresponding to those particles that went through the field without any gain or loss of corpuscles. This apparent randomness in their trajectory would show as a continuous track on the fluorescent screen. Of course, this totally unfounded mechanism relied on the supposition that all positive rays were necessarily H⁺ ions and that the different values of the charge-to-mass ratio were due to changing charge (for which the much smaller corpuscles would be responsible) rather than a variation in the mass of positive particles. At very low pressures, 'when there are very few ions in the gas, this continuous band stretching from the origin is replaced by discontinuous patches' (Thomson, 1907c, p. 10), which also seemed to reinforce his supposed mechanism of the continuous exchange of corpuscles on their way through the discharge tube. Furthermore, when he tried to observe the behaviour of positive rays in tubes filled with different gases, he did not find differences in the pattern of the fluorescence he observed (See Figure 4.3), even when, as he thought, he made sure that there was no hydrogen in the original tube (see Thomson (1907d)). His hope had been to obtain characteristic patterns depending of the atomic weight of the gas, but this he did not find except for the case of helium.

In many of these early experiments, especially those at relatively low pressures, J. J. Thomson found two clearly distinct patches, rather than one, with different maxima, the first corresponding to the charge-to-mass ratio   for hydrogen, the second to the same value for helium. The latter he reinterpreted as confirming the fact that helium ions, which had recently proved to be the alpha particles of radioactivity, played a central role in the structure of the atom. As for the characteristic parabolas he had initially expected from other gases, he hoped he might obtain them if he managed to obtain accurate results at very low pressures and high potentials. For the time being, he assumed that the positive rays inevitably broke up into the more elementary H^+ and alpha particles he was now observing. As Falconer showed, he never thought that the problem might come from the poor sensitivity of his willemite screen to heavy ions (Falconer, 1988, p. 281). What he did do, instead, was to make sure that there was no prior contamination of hydrogen in his experiments, which he (wrongly) thought he had successfully achieved.

The central tenet of his theory in the period 1907–1910 was what he called the neutral doublet. His observations also included a very strong luminescent spot corresponding to a large number of non-deflected, and therefore neutral, rays. His hypothesis was that the positive rays were produced *after* the cathode: in the strong field near the cathode, the gas molecules would release neutral doublets composed of a corpuscle and a positive element. The corpuscle might, in some cases, be set free from the doublet, producing an H^+ ion, which, in turn, would be deflected by the electric and magnetic fields in the tube in the way described above. In 1909, J. J. thought he had enough evidence to suggest that 'even at the start from the cathode the Kanalstrahlen include a large number of neutral doublets, if indeed they do not wholly consist of them', as he certainly hoped. From this, he speculated that 'these neutral doublets are very interesting, as they form an intermediate stage between the ion and the neutral molecule' (Thomson, 1909b, p. 828).

This intermediate state of matter would help explain the mechanism of ionization and, indirectly, the processes by which electricity was transmitted, while, at the same time, keeping his corpuscle-only theory of electrification. Furthermore, and although he did not make it explicit, this doublet model was reminiscent of the duality of charges involved in Faraday tubes (Falconer, 1988, p. 284). In any case, his results were still too provisional (and, as we shall see in the next chapter, totally wrong) to be a basis for any further speculation on the role of the positive electrification in the structure of the atom. From 1907 on, he did not make any refinement to his so-called 'plum-pudding' model: if he could not better understand the mechanism by which H^+ particles were formed, he had no arguments with which to modify his ideas on the structure of the atom.

At the height of his early research period on positive rays, J. J. Thomson was president of the British Association for the Advancement of Science, which, in 1909, met in Winnipeg, Canada. In his presidential address, J. J. described his work on positive rays as the latest step along a path that, starting with the discovery of X-rays, had revealed, in gases exposed to these rays, 'the presence of particles charged with electricity; some of these particles [being] charged with positive, others with negative electricity' (Thomson, 1909a, p. 10). The asymmetry between the two types of particles was, however, puzzling:

> We know a great deal about the negative electricity; what do we
> know about positive electricity? Is positive electricity molecular in
> structure? Is it made up into units, each unit carrying a charge equal
> in magnitude though opposite in sign to that carried by a corpuscle?
> Does, or does not, this unit differ, in size and physical properties, very
> widely from the corpuscle? (Thomson, 1909a, p. 12).

His experiments and those of other researchers at the Cavendish 'lead to the conclusion that the atoms of the different chemical elements contain definite units of positive as well as of negative electricity, and that the positive electricity, like the negative, is molecular in structure' (Thomson, 1909a, p. 13). Thomson was, however, too cautious to unequivocally affirm that he had isolated the particle of positive electricity, 'for we have to be on our guard against its being a much smaller body attached to the hydrogen atoms'. In other words, it might still be the case that positive electrification came in small particles, similar to the negative electrification, and that the bulk of the mass of the atom was due to a neutral mass.

That may explain why, from 1907, J. J. did not construct an alternative and consistent model for the atom. As with his discovery of the corpuscle, his first aim was to clarify the mechanisms and structure of electricity and only later moved to atomic structure. Since the nature of positive electrification was not yet clear, there was no basis on which to modify the existing model of the atom or propose a new one. In a laconic sentence in his presidential address, he acknowledged that 'since electrified particles can be studied with so much greater ease than unelectrified ones, ... we shall obtain a knowledge of the ultimate structure of electricity before we arrive at a corresponding degree of certainty with regard to the structure of matter' (Thomson, 1909a, p. 11). Furthermore:

> A knowledge of the mass and size of the two units of electricity, the
> positive and the negative, would give us the material for constructing
> what may be called a molecular theory of electricity, and would be
> a starting-point for a theory of the structure of matter; for the most
> natural view to take, as a provisional hypothesis is that matter is
> just a collection of positive and negative units of electricity, and that
> the forces which hold atoms and molecules together, the properties
> which differentiate one kind of matter from another, all have their
> origin in the electrical forces exerted by positive and negative units

of electricity grouped together in different ways in the atoms of the different elements. As it would seem that the units of positive and negative electricity are of very different sizes, we must regard matter as a mixture containing systems of very different types, one type corresponding to the small corpuscle, the other to the large positive unit (Thomson, 1909a, p. 14).

The so-called plum-pudding model, which was more the basis of a research programme than a real model, was, as it were, in quarantine. The properties of the atom had to depend, in ways not yet clear, also on the structure and arrangement of the carriers of positive, as well as negative, electricity.

4.6   Cambridge as a playground: George Paget Thomson

The visit to Canada was J. J.'s third trip across the ocean and the first one on which he took his son, now aged 17, with him. In the presidential address, G. P., who was soon to begin his undergraduate studies, heard his father describe the ideal training of a physicist in the following terms, which was not just a manifesto but also a rather accurate explanation of the path he had promoted to his son:

> ... the specialisation prevalent in schools often prevents students of science from acquiring sufficient knowledge of mathematics; it is true that most of those who study physics do some mathematics, but I hold that, in general, they do not do enough, and that they are not as efficient as they would be if they had a wider knowledge of that subject ... Two points of view are better than one, and the physicist who is also a mathematician possesses a most powerful instrument for scientific research with which many of the greatest discoveries have been made (Thomson, 1909a, p. 6).

George Paget Thomson belonged to that particular brand of British scientists with deep roots in Cambridge University. The son of 'Mr and Mrs J. J.', G. P. was born in 1892 at 6 Scroope Terrace, Cambridge, just a few minutes' walk from the Cavendish Laboratory. Until the age of 9, he was educated mainly by his mother: 'I never can be sufficiently grateful to my mother's teaching. Things were interesting and it was clear which had reasons and which had not' (G. P. Thomson, 1966, p. 13). Moreover, 'I was left-handed and my mother wisely allowed me to write that way, so I escaped the defect of stammering, then so common among the left-handed' (p. 4). In 1899, the Thomson family (at the time only Mr & Mrs J. J., and G. P.; his younger sister Joan, was born in 1902) moved to a detached house on West Road, where they stayed until 1920. From there, G. P. attended King's College Choir School and later the Perse School, both in Cambridge. His playground was at times the Backs, where he remembered 'playing at telephoning from some of [the trees] at a very early age' (p. 3), and most of his peers were the offspring of other well-established Cambridge families, since both schools were mainly day-schools.

At the time G. P. started going to school, J. J. was becoming internationally famous for his discovery of corpuscles, and, at the age of 14, he saw his father receive the Nobel Prize for his work on electric discharge. In his autobiographical notes, he recalled the office of his father as the *sancta sanctorum* of the house, into which he would make 'semi-legal invasions ... in the morning to borrow blue pencils' (G. P. Thomson, 1966, p. 3). In several accounts, J. J. is portrayed as a very absent-minded person (Rayleigh, 1942, Chapter 7). G. P. contributes to this idea, describing him as 'a much loved but inscrutable Jove, mostly in the Olympian clouds of his own thoughts' (G. P. Thomson, 1966, p. 16). He was distant, but not totally oblivious to his only son's needs. G. P. was passionate about making model sailing ships, a hobby that he maintained all his life (Moon, 1977, p. 531), and this may have helped in his decision to pursue a career in physics. The following anecdote is illustrative of the possible early influence his father had on his decision to become a physicist. The first models did not quite work:

> My mother, when consulted, spoke about centres of gravity as applied to tables and carts, but this was clearly different, so I took the case to the highest court, my father. He explained roughly how a ship differed from a cart for this purpose and said the stability could be calculated by mathematics and that this was done for real ships. I was much impressed, gave up the idea of a career in archaeology which had attracted me up till then and decided that mathematics and ships should be my future (G. P. Thomson, 1966, p. 11).

G. P. was given permission to get wood and materials from the Cavendish. The head mechanic of the laboratory, W. G. Pye, provided him with everything he needed and helped him in the construction of some of the biggest models, and J. H. Poynting provided him with a collection of working model guns for them. We can therefore imagine the young G. P. playing in the same laboratory where J. J. was working on cathode rays, C. T. R. Wilson was improving his cloud chamber, Rutherford was undertaking his first experimental work on X-rays, and Glazebrook was giving his last lectures in physics before resigning as Assistant Director of the Cavendish.

At the Perse School, G. P. was taught classics, mathematics, and sciences by teachers who were mostly graduates of Cambridge University. This is important since, as we saw in the first chapter, the particular way mathematics and physics were taught in the Cambridge Mathematical Tripos had spread to most public schools in Britain through high wranglers becoming schoolmasters. In Warwick's words, 'the success of the wrangler schoolmaster in reproducing the coaching system in schools and colleges throughout Britain is evident in the escalating levels of technical competence expected of Cambridge freshmen'. It was also significant that 'Wrangler masters also knew from firsthand experience how hard, for how long, and in what manner students needed to study in order to tackle difficult problems with the speed and confidence that would win them a Minor Scholarship to the university' (Warwick, 2003a, pp. 260–1).

However, G. P. Thomson had a very special pre-university education. Concerned that the Perse School might not be giving enough scientific training, his father arranged for him to receive private mathematical coaching by H. W. Turnbull (second wrangler in 1907), and to attend the brilliant lectures on introductory physics given by Alex Wood at the Cavendish: 'These were outstanding both in material and exposition, and impressed me greatly' (G. P. Thomson, 1966, p. 27). J. J. was happy with his son's choice of an academic career in science, and he was determined to help him pursue it: when the classics master of the Perse School suggested that G. P. was suitable for a career in classics, J. J. dismissed the suggestion and kept encouraging his son to follow in his footsteps and become a scientist. His advice was that G. P. should stick to the same path he himself had trodden in his youth, i.e., to study for the Mathematical Tripos and only later to move to experimental physics. 'He maintained that mathematics was a very important thing and you would learn physics somehow, roughly speaking' (G. P. Thomson, Oral Interview, AHQP, Tape T2, side 2). J. J. kept this opinion throughout his life. In his memoirs we can read: 'I am glad that I came under the older system, for I probably read much more mathematics than I should have done if I had taken my degree a few years later. I have found this of great value (*c'est le premier pas qui coûte*)' (Thomson, 1936, p. 39). Another relevant detail is that, during the summer between leaving King's and starting at the Perse school, J. J. also tutored his son:

> In the summer before going to the Perse I was taught for the first
> time by my father ... I tried not very successfully to teach myself
> the differential calculus, with greater success my father taught me
> elementary mechanics from Glazebrook's well known textbooks ...
> We also did some other mathematics. Whether by intent or otherwise
> he was not lavish in his explanations and I had to think harder than I

had ever done before, or indeed have often done since. One problem in continued fractions which he solved by what was virtually an application of finite differences bothered me for days and still sticks to my mind (G. P. Thomson, 1966, p. 20–1).

All these details help us understand the upbringing of someone who, at the time of becoming an undergraduate in the Cambridge Mathematical Tripos in 1910, had already deeply impressed upon his soul the particular ethos of this institution. As a result of his previous training, G. P. managed to take the exams for the Mathematical Tripos after two years, in the spring of 1912, when, at the time, most people took three years to finish the Mathematical Tripos. Since a Cambridge degree was only conferred after three years in residence, G. P. sat in on many lectures experimental physics, including those given by his father, and sat the exam for Part II of the Natural Science Tripos in the spring of 1913. Therefore, G. P.'s academic training included an unusual combination of advanced and fundamental mathematics together with a sound knowledge of experimental physics.

The teaching of physics in Cambridge in the early 1910s was still centred on the nature and properties of electricity and its interaction with matter. The world of Cambridge was, like most of the physics world of the time, still a world of ether. The two major Professors of Physics, J. J. Thomson and Joseph Larmor, had developed parallel theories in which the ether was one of the central entities. Larmor's (1900) book *Aether and Matter* and Thomson's (1903a) work *Conduction of Electricity through Matter* were the two major manifestations of Cambridge physics. There was almost nothing of the new quantum physics. 'I think,' G. P. said later in life, 'the quantum theory in my undergraduate days was something which was regarded by people with a good deal of reserve' (G. P. Thomson, Oral Interview, AHQP, Tape T2, side 2).

There is an interesting anecdote about his first days in college, which will prove relevant to the events to be described in Chapter 6. When his tutor recommended that he should attend lectures by the famous mathematician G. H. Hardy, G. P. already had strong opinions about mathematics and he challenged the advice. He had read an introduction to pure mathematics in which Hardy 'used in a simple form some of the more rigorous methods of proof till then little considered in British mathematical teaching though common abroad' (G. P. Thomson, 1966, p. 28). In later life, he regretted this decision, since it prevented him from following some of the mathematical developments of the new quantum mechanics of the mid-1920s, but he still justified it saying that 'the pleasure of mathematics lies less in any aesthetic appreciation of their beauty than in the sense of power over a range of ideas which ability to handle them mathematically brings with it' (p. 29). As we shall see, such an approach to mathematics created a strong divide between a generation of physicists for whom rationality was linked to mechanical models and those, like Dirac, for whom the beauty of mathematical symmetries was a strong element in the development of theories.

G. P.'s father taught a course called 'Electricity and matter', which contained much of his research after the discovery of the corpuscle. And father and son met regularly, every week, on Sunday for lunch and a walk in the fens. In these conversations, J. J. would recommend to his son which books to read, all of which he could, of course, find in his parent's house (G. P. Thomson, 1966, p. 35). Scientific collaboration between father and son was only a matter of time.

5

Father and son. Old and new physics

5.1   The nature of light

Is light a wave or a stream of particles? As we saw in Chapter 3, the sudden appearance of Roentgen rays on the stage in 1895 sparked J. J.'s interest in cathode rays, subsequently leading to his determination of the charge-to-mass ratio of corpuscles. The latter turned out to be the answer to his quest for a mechanism to explain electrical conductivity and the relationship between matter and electricity. Electricity revealed itself as corpuscular, as acting in fundamentally discrete units. Roentgen rays were instrumental in this atomization of charge, but they also opened up a more troublesome Pandora's box: that of the nature of light. Because it was very soon clear that Roentgen rays had properties similar to those of light at high frequencies, except that they did not spread uniformly in time: they behaved like pulses, like a stream of particles.

Since his early days as a Mathematical Tripos student, J. J. Thomson had lived in a world in which light was an electromagnetic wave in the ether. Certainly, many general histories of science talk about a permanent tension between the corpuscular and undulatory theories of light from the days of Newton until the advent of Einstein's quantum of light. But, in fact, light was, especially in late-nineteenth-century Cambridge, nothing but a wave. Hertz's experiments had consolidated what was a generally accepted theory of light into a 'virtual certainty' (Wheaton, 1983, p. 11). In this context, British physicists followed the path first forged by the Lucasian professor G. G. Stokes, explaining Roentgen rays in terms of impulses (Stokes, 1896), a hypothesis that seemed to be consistent with J. J.'s 1897 idea that cathode rays were electrified corpuscles. Consequently, in early 1898, Thomson published a paper in *Philosophical*

Magazine in which he put forward 'a theory of the connexion between cathode and Röntgen rays' (Thomson, 1898b). When a moving corpuscle suddenly came to a halt, some time was required for the change to propagate through the surrounding electric and magnetic fields, the further away from the corpuscle the more time it took. Such a change would be communicated in the form of a pulse generated by the stopping of the charged corpuscle in the electromagnetic field. Thomson showed mathematically that the width of such a pulse is proportional to the diameter of the corpuscle. 'The theory I wish to put forward', he said, 'is that the Röntgen rays are these thin pulses of electric and magnetic disturbance which are started when the small negatively charged particles which constitute the cathode rays are stopped' (p. 173). In 1903, he developed this idea more fully and in the more visual terms of his Faraday tubes:

> Let us consider the case of a charged point moving so slowly that the Faraday tubes are uniformly distributed, and suppose the point to be suddenly stopped, the effect of stopping the point will be that a pulse travels outwards from it …, but as the Faraday tubes have inertia they will until the pulse reaches them go on moving uniformly…, i.e. they will continue in the same state of motion as before the stoppage of the point…. Thus the stoppage of the charged particle is accompanied by the propagation outwards of a thin pulse of very intense electric and magnetic force; pulses produced in this way constitute, I believe, the Rontgen rays (Thomson, 1903a, p. 537–9).

The hypothesis that X-rays were regular pulses and not a periodic wave was, in 1903, generally accepted, both in Britain and in the Continent. The challenge that the first observation of the diffraction of X-rays had posed to the pulse hypothesis was, at least for a time, explained away by Sommerfeld: rather than imagining the pulse as the observational side of a set of otherwise-continuous waves (following Fourier's analysis), one could think of the pulse as a new form of light wave, there being some sort of continuity in the spectrum between the perfectly periodic waves of monochromatic light and the pulses forming X-rays (Wheaton, 1983, pp. 29–48). Also coming into the picture were the gamma-rays from radioactivity, which were increasingly understood as being analogous to X-rays.

The continuity between X-rays and ordinary light waves that Sommerfeld was suggesting found a speculative echo in J. J. Thomson. Now that he had a theory to account for the former, the latter surely *had* to be explained in similar terms. Thus, if one pulse of vibration on a Faraday tube came from the sudden stopping of a corpuscle, one could equally imagine that 'if a charged body were made to vibrate in such a way that its acceleration went through periodic changes, periodic waves of electric and magnetic force would travel out from the charged body' (Thomson, 1904a, p. 62). These would, by Maxwell's theory, be light waves. With this link between Faraday tubes and the propagation of light, J. J. introduced some sort of discreteness in the structure of light:

> The Faraday tubes stretching through the ether cannot be regarded as entirely filling it. They are rather to be looked upon as discrete threads embedded in a continuous ether, giving to the latter a fibrous structure; but if this is the case, then on the view we have taken of a wave of light the wave itself must have a structure, and the front of the wave, instead of being, as it were, uniformly illuminated, will be represented by a series of bright specks where the Faraday tubes cut the wave front (Thomson, 1904a, p. 63).

Around 1903, J. J. was concerned about the discrete ionization that X-rays seemed to produce. Already in 1896, Thomson was surprised by the existence of saturation in the ionization produced by Roentgen rays, a discovery that a Canadian student of his at the Cavendish, Robert McClung, corroborated in 1902. This meant that only a fraction of the molecules irradiated by X-rays were actually ionized, and that this ionization was independent of external factors such as the temperature of the irradiated substance. A discrete wave-front such as the one obtained using the image above of Faraday tubes could be the answer to this 'paradox of quantity' in ionization (Wheaton, 1983, chapter 4).

J. J. came back to this fibrous structure of the ether in 1907, after briefly embracing the popular 'triggering hypothesis' first suggested by Lenard to explain the photoelectric effect. According to the latter, the fact that the emission of corpuscles by incident light was dependent on its frequency and not on its intensity could be explained if the energy with which corpuscles were emitted pre-existed in the atoms, the role of the incident beam of light being only to trigger their release. 'On this view,' Thomson wrote, '... the rays act as detonators, causing some of the atoms on which they fall to explode, and the energy of the corpuscle is derived from the energy liberated by this explosion' (Thomson, 1905c, p. 588). That would relate photoelectricity to radioactivity, both being a release of energy due to some internal rearrangement of the corpuscles in the atom. Not surprisingly, J. J. related this to the internal structure of the atom, about which he speculated much around that time:

> We can, however, easily conceive an atom constructed in such a way that before the internal energy had diminished sufficiently to appreciably alter many of its properties, the atom would become

unstable and explode, breaking up into atoms of elements of a different kind. Suppose, for example, that the atom consists of a number of corpuscles arranged in layers on the surfaces of concentric shells, and that the loss of internal energy by the atom as mainly due to the loss of kinetic energy by those corpuscles in the outer layer, this will hardly affect the times of vibrations of the corpuscles inside, while the outer layers may lose such a large amount of energy that their configuration becomes unstable, and the corpuscles in the outer layers rearrange themselves: in doing this, such a large amount of kinetic energy may be liberated that the atom explodes and breaks up into atoms of different kinds. Thus, in a case of this kind we should have the atom losing internal energy and yet as long as it remained intact the great majority of its periods of vibration would be unaltered, and the atom would explode before the change in its internal energy was sufficient to appreciably affect the great majority of its properties (Thomson, 1905c, p. 590).

Although the triggering hypothesis as an explanation of the photoelectric effect was very popular among scientists until 1911 (Wheaton, 1983, p. 75), J. J. abandoned it in 1907, for he found a way to account for both ionization and the photoelectric effect using his model of pulses travelling in Faraday tubes. The way this happened is a good example of the speed at which J. J.'s speculative mind could change to account for the latest experimental input. At the end of June, he presented a paper by a research student of his, P. D. Innes, to the *Proceedings of the Royal Society* giving further evidence for the explosion theory (Innes, 1907). Innes showed that when different metals were irradiated with Roentgen rays, the velocity of the emitted photoelectrons[1] was independent of the energy of the incident radiation. Encouraged by J. J.'s explosion theory, Innes argued that the speed of the emitted electrons could largely be explained as the result of an accumulation of internal energy in the corpuscles of the atoms and subsequent explosion. By October, however, J. J. had changed his mind and gone back to his explanation in terms of a structure of light after hearing about the experiments of Erich Ladenburg, a Berlin-based physicist (Ladenburg, 1907).

Innes had also shown that the velocity of the photoelectrons emitted by different metals was dependent on the atomic number of the atoms, but he did not pay direct attention to the range of velocities from a given metal.

[1] This paper by a research student of J. J. Thomson shows the terminological fuss still present at the Cavendish in 1907: the terms electrons, corpuscles, photoparticles, and β-corpuscles are used interchangeably throughout the paper.

Ladenburg's experiments with ultraviolet light did look into this variable and concluded that the velocity of the emitted electrons varied continuously with the frequency of the incident light. For Thomson this was a problem: if the role of the incident radiation was to trigger an internal explosion in the atom, one should expect some frequencies to be particularly efficiently generated, i.e., those resonating with the internal vibrations of the corpuscles. But since that was not the case, J. J. interpreted Ladenburg's experiments as implying that every atom had a large number of internal vibrations among its corpuscles so that each incident frequency found a resonating mode within the atom. At a time when the number of corpuscles had already been cut down to the order of the atomic number, this was 'improbable'. 'It seems more reasonable,' he said, 'to suppose that the velocity is imparted by the light, and yet... the velocity is independent of the intensity of the light' (Thomson, 1907e, p. 422). His suggestion was that these results could only make sense if 'a wave of light is not a continuous structure, but that its energy is concentrated in units', as he had already put forward in 1903.

The image he used to visualize the discrete structure of light followed the thread of his 1903 theory. He supposed that 'the ether has disseminated through it discrete lines of electric force and that these are in a state of tension and that light consists of transverse vibrations, Röntgen rays of pulses, travelling along these lines' (Thomson, 1907e, p. 421). The energy of the wave would be concentrated in these pulses, thus giving a discrete appearance to the wave-front when traversing a black screen: 'the energy of the wave is thus collected into isolated regions, these regions being the portions of the lines of force occupied by the pulses or wave motion'. The effect would be, of course, very similar to that given by what he called 'the old emission theory' that spoke of corpuscles of light. The independence from intensity was explained in the following terms: 'if we consider light falling on a metal plate, if we increase the distance of the source of light', and considering spherical symmetry from the source, 'we shall diminish the number of these different bundles or units falling on a given area of the metal, but we shall not diminish the energy in the individual units'.

The latter would explain why the energy of the emitted particle did not depend on the intensity of the incident light. But J. J. had now to explain Ladenburg's finding that the speed of the photoparticles was dependent on the frequency of the light with which it was irradiated. He faced this problem by moving a step backwards and incorporating into the picture his theory of the formation of Roentgen rays (which, by 'analogy', would be valid for other forms of light): when cathode rays suddenly stopped, they would emit Roentgen rays, with the more rapidly moving (thus, more energetic) cathode particles producing thinner pulses (thus, higher frequencies). Thus, the energy present in each unit of light would depend on its frequency, which is what Ladenburg's experiments showed.

Although, ten years later, Robert Millikan saw this theory as almost equivalent to Einstein's 1905 corpuscular theory of light (Millikan, 1917, pp. 221–3), it is clear from Thomson's words that his structured light is perfectly within the bounds of ether physics. It is the physicality of Faraday tubes which allows for this structure of light:

> Thus the structure of the light would be of an exceedingly coarse character, and could perhaps best be pictured by supposing the particles on the old emission theory replaced by isolated transverse disturbances along the lines of force. The greater the frequency of the light the greater is the energy in each unit, so that if it requires a definite amount of energy to liberate a corpuscle from a molecule of a gas, light whose wave length exceeds a particular value, which may depend on the nature of the gas, will be unable to ionize the gas, for then the energy per unit will fall below the value required to ionize the gas (Thomson, 1907e, p. 423).

As is well known, the tension between corpuscular and undulatory theories of light persisted until the general acceptance of the Einstein's quantum of light and the formulation of a generalized principle of wave-particle duality, both in the mid-1920s (Stuewer, 1975). In the meantime, physicists had to come to terms in the best way they could with what J. J. famously called a 'battle between a tiger and a shark' (Thomson, 1925, p. 15).

5.2   The early theory of the quantum

In the aforementioned presidential address of 1909 to the British Association for the Advancement of Science in Winnipeg, J. J. felt there was a need to stress the reality of the ether:

> The ether is not a fantastic creation of the speculative philosopher; it is as essential to us as the air we breathe. For we must remember that we on this earth are not living on our own resources; we are dependent from minute to minute upon what we are getting from the sun, and the gifts of the sun are conveyed to us by the ether. It is to the sun that we owe not merely night and day, springtime and harvest, but it is the energy of the sun, stored up in coal, in waterfalls, in food, that practically does all the work of the world (Thomson, 1909a, p. 15).

We find this statement in the context of a long survey on the *status* of contemporary physics: the nature of light, the interaction between electricity and matter, and the origin of energy in radioactivity; they all seem to point towards the ether, in spite of its seemingly contradictory properties. 'The study of this all-pervading substance,' he says, 'is perhaps the most fascinating and important duty of the physicist' (Thomson, 1909a, p. 15), as opposed to the chemist, whose main research topic should be matter and its combinations. This clear defence of the reality of the ether acts as the preamble to his first public mention of the 'very remarkable series of investigations on the Thermodynamics of Radiation' by Planck, which has 'lately received a great deal of support' (p. 21).

This brings us to a new question, one that remains cogent through the rest of this book: what was J. J. Thomson's attitude towards the new quantum theory emerging mainly in the German-speaking world? As we just saw, J. J. was not unaware of the fact that some exchanges of energy between radiation and matter presented discrete aspects, but he was continually bringing forward some ad hoc mechanism to account for such *apparent* discreteness. The theory of the quantum that was gaining momentum on the Continent took the quantum of energy as the primary explanatory tool without speculating on what kind of internal mechanism caused energy to be transferred in multiples of Plank's constant h. The quantum was only loosely conceptualized, but this did not prevent it from gaining popularity among an increasing number of physicists, especially after h was seen to explain not only the radiation of a black body but also the specific heat of solids. A most radical interpretation of the quantum held that h was not only a phenomenological feature in the interchange of energy, but that it was moreover the atomic unit of energy itself. This latter interpretation was linked to Einstein's 1905 suggestion of the quantum of light and had only a few early supporters like Johannes Stark.

As Russell McCormmach argued long ago, J. J. probably studied Planck's theory from his 1906 book on heat radiation (Planck, 1906), and understood it as an alternative physical model explaining the structure of light, more along the lines of Einstein and Stark than that of Planck himself (McCormmach, 1967, pp. 374–8). According to Thomson, Planck had deduced the second law of thermodynamics using the quantum hypothesis, but that could not be a good strategy: 'if it were a legitimate deduction,' he wrote, 'it would appear that only a particular type of mechanism for the vibrators which give out light and absorbers which absorb it could be in accordance with [the Second Law of Thermodynamics]' (Thomson, 1909a, p. 21). But this *obviously* had to be wrong since 'if this were so, then, regarding the universe as a collection of machines all obeying the laws of dynamics, the Second Law of Thermodynamics would only be true for a particular kind of machine'. In other words, J. J. Thomson could certainly not agree to a theory in which the transfer of energy was in discrete units not as a result of the nature of the mechanism (like in his Faraday tubes model) but as an a priori imposition to the model. The quantum had to be a consequence, not a precondition.

J. J. encouraged a number of people at the Cavendish to check for evidence of discreteness in phenomena like interference or correlation at very low intensities otherwise perfectly explained by the wave theory of light. Among them, Norman R. Campbell, a research assistant at the Cavendish and fellow of Trinity College, was the only one seriously convinced of the validity of Einstein's quantum of light and the need to do away with the ether. Sadly for him, and 'contrary to the hopes of the author, no evidence has been produced against the "spherical wave" theory' (Campbell, 1910, p. 521), since his experimental setup produced too many fluctuations. In any case, J. J. Thomson was increasingly nervous in the face of what some, including Campbell, started to regard as the 'modern theory', free of ether and continuity (Campbell, 1909, p. 117), and he moved a step forward in his search to accommodate the discrete phenomena of light in a continuous ether-filled world.

He now thought of reducing the number of Faraday tubes originating from a corpuscle to one. Following the tradition that the electric field spreads out from a charged body in all directions, J. J. had so far imagined a large number of Faraday tubes starting from one corpuscle and dispersing with spherical symmetry into space. But now he decided to regard this uniformity of the field in all directions as a statistical measure stemming from the fact that most work on electricity was done with bodies containing a large number of corpuscles, 'the result [being] the same whether each individual field is uniformly distributed in all directions or is confined within a small solid angle' (Thomson, 1910a, p. 302). What he got from this was that 'the electric field due to a number of corpuscles is a mosaic, as it were, made up of a number of detached fields. The electric field itself, as well as the electric charges in it, being molecular in constitution'. As in his previous model, radiation originated in the sudden stopping of a corpuscle and the transmission of the corresponding kick along the Faraday tube. By contrast, however, energy did not spread in all directions but only in one: the direction corresponding to the one and only Faraday tube. His theory more and more resembled the old emission theory, treating each kick in the Faraday tube as a unit of light that could interact with other units in the way particles do when they collide, but would 'be in agreement with the undulatory theory in supposing that the electrical disturbance whose propagation constitutes light is a vector quantity' (p. 311), i.e., that the electromagnetic field travels perpendicular to the direction of the kick. (Interestingly, what

J. J. referred to as the undulatory theory was Maxwell's theory of light, not so much the fact that light was a wave, but the fact that light was the propagation of an electromagnetic disturbance.)

As for phenomena like interference, he thought that, unlike a purely corpuscular theory of light, his theory could also account for it. In his view, one could get interference if a large number of Faraday tubes with related frequencies in their fluctuations went through a slit. And this might be possible, taking into account that, although each Faraday tube was originating in one corpuscle only, one could easily imagine that corpuscles close to each other would have movements of related frequencies: 'For consider a corpuscle vibrating in a definite period; in its neighbourhood there will be many other systems having the same time of vibration, and the vibrations of these will be excited by resonance and will be in phase relation with the primary vibration' (Thomson, 1910a, p. 311-2). Even though, as usual, J. J. basically stayed at a qualitative level, his model appeared to be far superior to the quantum hypothesis, since the latter could not at all explain interference phenomena. We should perhaps remember that, partly because of this, few people were convinced by Einstein's quantum of light in the 1910s and that, in this respect, J. J. was not alone in opposing it.

The interaction between radiation and matter, which had been the origin of Planck's hypothesis, was a different matter. The early quantum theory was gaining in popularity since it was proving successful in explaining an increasing array of phenomena. In 1911, the first Solvay conference met in Brussels, becoming the first international meeting dealing with the question of the quantum. J. J. Thomson was invited but did not attend possibly due in part to disagreement with the emphasis on the quantum and in part to his reluctance to travel to places where English was not the *lingua franca*. That he did not like the favour Planck's theory was gaining is clear in a number of his papers between 1911 and 1914, in which he directly addressed the theory of the quanta and its limitations. In a 1910 paper where he discussed a theory of radiation based on the collision of corpuscles with molecules, he admitted that 'there are many phenomena which can be interpreted as indicating that the energy in radiation is made up of definite units, and that these units are indivisible' (Thomson, 1910b, p. 243). But this was no reason for such a theory to be the right one since, he asked, 'why should a unit of light when passing over a corpuscle be obliged to communicate to it either the whole of its energy or none at all?' (p. 244). To address this problem, he brought in an element from his research on positive rays, the neutral doublets, as a possible substructure of the atom. He imagined that, in the atom, one might have a number of polarized neutral doublets with one corpuscle orbiting around each of the doublets. When light of a certain frequency traversed matter, only those corpuscles with frequencies that resonated with that of the incident light would be released. That did not mean that the rest of the corpuscles were not affected: the incident light would twist their rotation movements around their doublet, but would not eject them from their orbits, thus having no global effect.

In 1913, he proposed a more radical solution also relating the inner structure of the atom to the challenges of quantum phenomena. In a series of lectures at the Royal Institution early that year, he moved a step forward and argued that 'we cannot assume that the forces due to the charges of electricity inside the atom are of exactly the same character as those given by the ordinary laws of Electrostatics' (Thomson, 1913a, p. 793). And he suggested that corpuscles inside the atom were subject to two kinds of forces: an attractive one, proportional to the square of the distance, and a repulsive one, proportional to the cube of the distance. Capitalizing on his earlier suggestion that every corpuscle was the origin of only one Faraday tube, he now assumed rather that each corpuscle was trapped in one tube of force, not entering 'at this stage into any consideration as to the origin of this force; we shall simply postulate its existence' (p. 794). The atomic corpuscle could oscillate in the direction of the tube, but needed a minimum amount of energy to move transversely and quit the tube. This minimal energy would coincide with multiples of Planck's constant. Once again, his main point was to emphasize that one need not assume that 'radiant energy is molecular in structure', but that the same results could be obtained 'if the mechanism in the atom by which the radiant energy is transformed to kinetic energy is such as to require the transference to the mechanism of a definite amount of energy' (p. 792). However, the mechanisms J. J. was putting forward were more and more ad hoc and incapable of giving a consistent picture of the structure of the atom.

Perhaps the clearest and most explicit attack on Planck's theory was his 1912 communication in the *Proceedings of the Cambridge Philosophical Society*, 'The unit theory of light' (Thomson, 1912a). There, he tried to show that every single phenomenon accounted for by the theory of the quantum, both in the absorption and the emission of radiation, could be equally explained by supposing some intra-atomic mechanism in which there was a threshold energy for the ejection of a corpuscle from an atom. That would explain why energy was absorbed in apparently discrete units without having to assume that energy had itself a molecular structure. While both theories were equally explanatory of the experimental data, Thomson's was superior to Planck's in that the former preserved the continuity in optical phenomena, phenomena for which the latter had no explanation, 'and in the present state of the subject at any rate is a hindrance and not a help' (p. 643). Thomson could thus preserve the undulatory aspects of radiation precisely because, even though internal mechanisms in the atom produced discrete emission, these quanta were not essential and could, through their interaction with matter, be subdivided until they approached a 'nearly continuous' distribution (p. 650).

5.3   Britain and the quanta in 1913

The best picture of the situation of quantum theory in Thomson's and many other British physicists' minds in 1913 can be garnered from the British Association meeting in Birmingham. James Jeans, who had recently converted to the theory of the quantum and was one of only two British physicists present at the 1911 Solvay meeting, took on the task of explaining and defending the theory of the quantum to a reluctant audience. Jeans had been second wrangler in 1898, and he and G. H. Hardy were the first to attempt the Cambridge Mathematical Tripos in only two years – and not in the usual three years – after which he was appointed fellow and lecturer in Trinity College (Milne, 1952). After graduating, he worked on radiation theory and statistical mechanics, producing his first book, *The Dynamical Theory of Gases* (Jeans, 1904), and contributed to the formulation of what we now call the Rayleigh–Jeans law for the distribution of radiation of a black body, which was derived using the equipartition of energy. His consistent failure to describe the experimental energy distribution of black-body radiation using classical arguments did not lead Jeans, at first, to accept Planck's hypothesis, but to search for alternative mechanisms to explain the experimental law. Faithful to the equipartition principle, central in statistical mechanics, Jeans was at first willing to challenge Planck's law on the basis that real, physical equilibrium was impossible in a black body. But, by 1910, he had changed his mind, forced by the experimental success of Planck's law as well as by the theoretical proof that this law could be obtained *only* with the assumption of quanta (Hudson, 1989). Another recent convert, Henri Poincaré, also developed a very detailed demonstration of the sufficiency and necessity of the hypothesis of quanta to obtain Planck's law in 1912, just after the first Solvay conference.

The discussions at the Birmingham meeting 'made it abundantly clear that the quantum-theory is far from being regarded as inevitable yet by many of the English school of physicists' (Jeans, 1914, p. 23), and when Jeans published his *Report on Radiation and the Quantum-Theory* a few months later, he fairly included full references to the criticisms by Thomson, Larmor, and others (Navarro, 2012). Addressing Thomson's most recent models, Jeans noted that his suggestion of repulsive forces varying according to the law of the inverse cube 'is not a condition which can be easily reconciled with what is known about the structure of atoms and the motion of electrons' (Jeans, 1914, p. 28). Furthermore, one could not accept discreteness in the absorption and emission of radiation without full commitment to what Jeans here called the 'quantum-theory' as the guiding principle. This was precisely what J. J. and many others were still doing: accepting a certain amount of phenomenological discreteness without embracing it as a principle of nature or as an ultimate explanatory tool. Incidentally, the 1913 British Association meeting started with a presidential address given by Oliver Lodge on 'Continuity', a manifesto in favour of the real existence of the ether and its essentially continuous nature, and against the theories of relativity and the quantum (Lodge, 1914). A generational gap was clearly opening up.

Jeans' attitude, however, was also symptomatic of a particularly British way of accepting the quantum theory. For him, the main problem was not that the quantum theory was, so far, limited in its applicability, but that 'even if the complete set of equations were known, it might be no easy task to *give a physical interpretation of them, or to imagine the mechanism* from which they originate' (Jeans, 1914, p. 79, my emphasis). I have emphasized the last sentence because for him, as for most physicists of the Cambridge school, intelligibility involved the possibility of imagining a mechanism, even if only heuristically, that accounts for the observed phenomena. But when faced with the quantum, any 'attempt to imagine a universe in which action is atomic leads the mind into a state of hopeless confusion' (pp. 79–80).

The last chapter of the *Report* finishes with a discussion on the reality of the ether, acknowledging that, in this respect, Continental and British physicists fought on different – opposed – sides. Jeans seemed to cling to the reality of the ether, but he relegated it to second place, the real stumbling block being the contradiction between discrete and continuous theories, each valid for different radiation phenomena. And, with this, the last pages of the book convey a certain amount of pessimism for the status quo of physics. In a free translation from Poincaré's *Dernière Pensées*, he wrote:

> It is impossible at present to predict the final issue. Will some entirely different solution be found? Or will the advocates of the new theory succeed in removing the obstacles which prevent us accepting it without reserve? Is discontinuity destined to reign over the physical universe, and will its triumph be final? Or will it finally be recognized that this continuity is only apparent, and a disguise for a series of continuous processes? … Any attempt at present to give a judgement on these questions would be a waste of paper and ink (Jeans, 1914, p. 90).

While most of the *Report* was an active exercise in convincing the reader of the inevitability of the quantum hypothesis and its successes, these last pages brought that optimism back to the earth by pointing out the difficulties in interpreting the quantum theory. But this was done in a particular way: these last sentences can be interpreted as a way of encouraging British physicists to embrace the theory without a priori rejecting it on the grounds that it was not 'physical', i.e., mechanical. Furthermore, the fact that these considerations appeared only at the end of the book, as a separate chapter, may indicate that, from Jeans' point of view, one could and should accept the quantum theory without having a complete notion of its ultimate physical meaning. Partly following the problem-solving tradition of the Cambridge Mathematical Tripos pedagogy, Jeans was more concerned with proving that the quantum theory solved specific problems than attempting an overall criqitue on metaphysical grounds.

5.4   A father–son collaboration

Between 1901 and 1914, more than 200 researchers worked at the Cavendish, producing almost 400 research papers (Kim, 2002, p. 164). Many of these were the result of suggestions by Thomson to young scientists in search of a research topic at the forefront of physics. In a way, J. J. partly turned the Cavendish into an institution mass-producing results pertaining to all sorts of radiation: monochromatic light, X-rays, radioactivity, photoelectricity, etc. When J. J. had an idea about the nature of radiation, the process of ionization, or a theory of conduction, he could easily find a research student eager to put such idea to the test, enabling him to be very prolific in the production of an ever-increasing number of theoretical models. Interestingly, however, J. J. himself performed almost no experiments related to the structure of light, X-rays or gamma-rays. His personal experimental project continued to be positive rays throughout, from 1910, with the collaboration of the new research assistant at the Cavendish, Francis Aston, and from 1913, with his son G. P.

J. J.'s work on positive rays had, by 1910, come to a standstill. His experimental set-up was too sensitive to changes in physical variables, especially at low pressures, and his results were so full of errors that no consistent theory could be developed. As we saw in the previous chapter, there was no way to decide if the primary emission during ionization consisted of positive rays or his supposed neutral doublets. Moreover, the research student with whom he had worked on this topic since 1907, George Kaye, left for the National Physical Laboratory. Francis Aston, who had studied chemistry and physics in Birmingham under Poynting, and then became a researcher at that university, filled the position. Unlike J. J., Aston was a highly skilled experimentalist, dexterous in glass blowing, and always eager for better and more accurate experimental results. Aston's arrival brought about a change in the pace and the overall aim of J. J.'s project on positive rays (Hughes, 2003, 2009).

Aston suggested that, in order to get controllable discharges and, consequently, accurate results at low pressures, they should work with larger vessels, following an idea that J. J. himself had suggested long before in his *Notes on Recent Researches* but never implemented (Thomson, 1893, p. 90). This modification, with bulbs of up to eleven-litre capacity, dramatically changed the results at low pressures: 'phases of the phenomena of the positive rays come to light which are absent or inconspicuous at higher pressures' (Thomson, 1910c, p. 752). He could now clearly observe on the screen the characteristic parabolas for each gas contained in the tube: each parabola corresponded to a particular charge-to-mass ratio, as he had predicted in 1907. Another radical change came with the use of photographic plates rather than a willemite fluorescent screen to record the parabolas. And they also managed to get a better vacuum pump for the deflection tube, so as to minimize the neutralization process. With these changes and the new results, he abandoned the idea that neutral doublets might be at the origin of the emission of positive rays, and he slowly came to the realization that his ubiquitous H^+ was due to hydrogen contamination and not to the supposed existence of a fundamental unit of positive electrification. But instead of explicitly admitting the experimental error, J. J. avoided any mention of it, took what could be saved, and turned his research on positive rays into a totally different project (Falconer, 1988, p. 303).

Already in the first paper he published after these modifications were implemented, in September 1910, he suggested a possible use for this new technique: 'I think this effect may furnish a valuable means of analysing the gases in the tube and determining their atomic weight' (Thomson, 1910c, p. 758). And after a few months' work, this turned into the main goal of his research on positive rays, as the title of his public lecture at the Royal Institution, in April 1911, makes clear: 'A new method of chemical analysis' (Thomson, 1911). The parabola method would constitute an alternative method to spectroscopic chemical analysis, one that 'is even more delicate than that of spectrum analysis, for by it we can detect the presence of quantities of a foreign gas too minute to produce any indication in the spectroscope' (p. 3). Other benefits of this young 'but sufficiently developed' method over spectroscopy included the very small quantity of material needed for a complete chemical analysis (no more than 1/100 of a milligram), and more importantly, the fact that it could not only identify new elements or new ionic forms of a given substance but also give their atomic mass without further analysis, since each positive element would

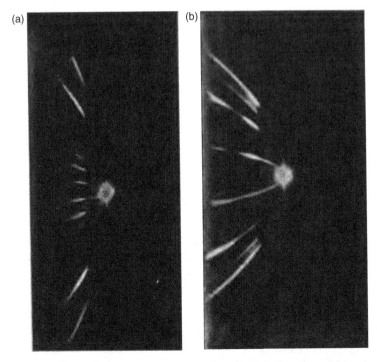

Figure 5.1. Images of the parabolas obtained for (a) nitrogen and (b) carbon monoxide (Thomson, 1911, figures 1 and 3). Courtesy: Royal Institution of Great Britain.

give a separate parabola. Finally, one need not have a pure substance in order to analyse its constitution, since any impurity would trace out an altogether different parabola. And, by way of illustration, he showed some of the clearest pictures he had so far obtained (Figure 5.1).

His papers from 1910 onwards contain systematic lists of observed atoms and molecules, many of which had never been isolated. Thus, for instance, J. J. obtained evidence for the carbon radicals CH, CH_2, and CH_3. He also began to detect multiply charged ions, the most striking example being mercury, which was present with one, two, three, … and up to eight units of electricity. That posed a new problem that led him to modify his views on the production of positive rays. So far, J. J. had largely thought that positive rays were the result of ionization: cathode rays in the ionization tube would collide with the molecules of the gas, trigger the release of one corpuscle, and leave a positively charged ion of the gas in the tube. That explanation was plausible for singly charged, or perhaps even for doubly charged ions, but it seemed totally implausible for higher charges like the ones he was now obtaining. His suggestion was that 'in the discharge-tube there are two, and only two, kinds of ionization; in one of these kinds the mercury atom loses 1 corpuscle, while in the other kind it loses 8' (Thomson, 1912c, p. 670). Atoms from the second ionization could regain a few corpuscles, thus giving rise to the whole range of ions as secondary phenomena. The important thing here was that, for the first time since the discovery of the corpuscle, J. J. thought of an ionization process not caused by them. In the first ionization, the fast corpuscles from cathode rays 'penetrate into the atom and come into collision with the corpuscles inside it individually, the collision in favourable cases causing the corpuscle struck to escape from the atom'. In the second case, however, 'we suppose that the mercury atom is struck by a rapidly moving atom and not by a corpuscle; after the collision the mercury atom starts off with a very considerable velocity, which at first is not shared by the corpuscles inside it'. As a result, 'if there were eight corpuscles in the mercury atom connected with about the same firmness to the atom, the result … might be the detachment of the set of eight leaving the atom with a charge of 8 units of positive electricity' (p. 671).

J. J. extended his dual method of ionization to all elements, since he had seen ions of one and two charges in most elements, with the exception of hydrogen that was only present in the H^+ form. For the rest, he now inferred that the doubly charged ion was always the result of atomic collisions, and not, as previously thought, due to the collision of corpuscles with the atom. And as supporting evidence, he mentioned the case of helium and its relationship with the alpha particles from radioactivity: 'in the vacuum-tube the helium atom occurs with both single and double charges, whilst as an α particle it always seems to have two charges, suggesting that the process by which the α particle acquires its charge is analogous to the process by which multiply-charged atoms are produced in the discharge-tube' (Thomson, 1912c, p. 672). With this, he related the formation of multiply charged ions to the emission of alpha particles in radioactive materials, but, more importantly, he was slowly giving a new status to the positive part of the atom as a substantial unit that could have an independent life without all its valence corpuscles. As we saw in Section 5.2 above, his 1913 paper 'On the structure of the atom' *indirectly* contains the idea of a unit of positive electrification in the centre of the atom, with the external corpuscles linked to it by the mediation of Faraday tubes through which radial attractive and repulsive forces are transmitted.[2]

But perhaps the result that fascinated him most was the detection 'of some compounds that have not hitherto been detected by chemical methods'

---

[2] I say indirectly because in that paper, as in many other contributions around those years, the emphasis was on the forces that act upon the corpuscles rather than the overall structure of the atom.

(Thomson, 1912b, p. 240). His favourite and, so he thought, most relevant discovery was that of what he called X_3, an atom with a charge-to-mass ratio three times that of H^+. This ratio could be due to a carbon atom with four charges or to a singly charged molecule containing three hydrogen atoms. Since this parabola was obtained in experimental situations free from the presence of any carbon compound, J. J. decided that it had to be the new H_3^+ molecule, even though systematic analysis of the new substance gave no clear results. As Falconer convincingly demonstrated, this interpretation was very appealing in the context of atomic structure (Falconer, 1988, pp. 305–7): around 1913, J. J. was toying with the idea that the core of the atom increased in units of four, i.e., by addition of alpha particles, which would explain why positive radioactivity was emitted only in these units. In his usual loose manner, he started from the *fact* that all atoms up to mass 40, except, of course, hydrogen, seemed to fall into the series $4n$ or $4n + 3$ (a *fact*, that by the way, was not true for beryllium or nitrogen). In this case, H_3 might prove to be the first element in the second series. Less appealing, for him was, a new parabola associated with atomic weight 22 associated Ne^{20}, and which he explained away as possibly due to the ion NeH_2^+. It was Aston who focused on this latter line after being awarded the Maxwell scholarship in 1913, thus obtaining intellectual independence from J. J. The story continues with Aston's interpretation of this line as an isotope of neon, the first non-radioactive isotope to be discovered in the laboratory, and the first of a series of achievements with his modified experimental device: the mass spectrograph (Falconer, 1988; Hughes, 2003, 2009). J. J.'s researches with positive rays continued now with his son as collaborator.

In the summer of 1913, G. P. Thomson graduated and decided to start research work at the Cavendish, after being offered a research fellowship at Corpus Christi College. Wanting to pursue a scientific career, he needed to find a topic to develop as his own research and a place to work on it. His decision can be seen as slightly conservative: to stay in the family, at the Cavendish, with his father as supervisor. This choice, although not unusual, and certainly not a bad one, deserves some reflection. In 1913, the Cavendish was no longer the only vibrant place for physics in Britain as it had certainly been ten years earlier: while, in the early 1900s, the Cavendish was virtually the only destination for 1851 Exhibition Scholars, in the early 1910s, an increasing number of them decided to move to other British research centres (Kim, 2002, pp. 169–74). New laboratories, such as those in Manchester and Leeds, had become established, with research programmes more on the cutting edge of science (Hughes, 2005). Perhaps the best-known example of disappointment with the Cavendish was Niels Bohr. Full of energy and with his recent Ph.D. dissertation on a theory of electrons, Bohr was eager to meet Thomson as the father figure of the electron, only to find a person doing experiments in the same old way that 15 years earlier had led him to the formulation of the corpuscle hypothesis, and quite reluctant to introduce conceptual changes into his world-view. Bohr decided to move to Manchester, where he eventually developed his quantum model of the atom.

If those from outside Cambridge were starting to look at other possible locations to do their research, some Cambridge graduates were doing the same. C. G. Darwin, a life-long friend of G. P. Thomson and, like him, born and raised in that provincial town, decided to move to Manchester, aware of the subtle, but increasing, limitations of the Cavendish (G. P. Thomson, 1963). Another example was William Bragg, who had graduated in 1912 and started his own X-ray research project and who considered the laboratory a 'sad place' (Hunter, 2004, p. 21). With this, I want to illustrate that the Cavendish was certainly not the only possible exciting destination for a Cambridge graduate of G. P.'s generation, which may lead us to infer that he considered it a positive thing to stay intellectually close to his father.

The status quo of positive-ray research at the time G. P. joined his father is reflected in the book *Rays of Positive Electricity and their Application to Chemical Analysis* that J. J. published in 1913, and which contains basically the work already published in previous papers. In putting this book together, J. J. wanted to promote his method of analysis using positive rays to other physicists and chemists: 'I feel sure that there are many problems in Chemistry which could be solved with far greater ease by this than by any other method' (Thomson, 1913b). These were the discovery of new substances and transitory combinations (like his finding of previously unknown radicals), the immediate determination of whether a substance was diatomic or monatomic (the former giving two parabolas, the latter only one), the measurement of the proportions of different substances in a mixed gas, or the possibility of different electrifications of a given substance. Furthermore, he thought that much of the information he was accumulating through his work on positive rays was relevant to understanding the problem of chemical bonding, which we shall consider in the last section of this chapter.

As we shall see in the next section, war broke out in 1914 and the paths of J. J. and G. P. Thomson, father and son, went in different directions. G. P., like most young British scientists, enlisted as an ordinary soldier. After a few months in France, he was sent back to England, to work as a scientist at the Royal Aircraft Factory on problems concerning aerodynamics and the building of aeroplanes. This provided him with the possibility of developing that aspect of physics that had, as a child, triggered his interest in science, as well as removing him from the theoretical world of Cambridge and the research projects of the Cavendish.

The thorough knowledge of aerodynamics he acquired during the war resulted in a book on the subject written in 1919 (G.P. Thomson, 1920a).

As the war came to an end, J. J. became Master of Trinity College, and agreed to resign from his position as director of the Cavendish, a position to which Rutherford was appointed, bringing with him the school of radioactive research he had built in Manchester (Wilson, 1983, pp. 406–13). In spite of all these changes, when the war ended, G. P. decided to continue the research project on positive rays with his father. Later in life, G. P. justified this move, saying that 'the positive ray … was a very big thing in the Cavendish' (G. P. Thomson, AHQP, T2, 2, 12). It was certainly a 'big thing' for his father who, in 1920, was preparing a secondly highly revised, edition of his book on positive rays. And it was a 'big thing' in the mind of Aston who, while also working on the construction of aeroplanes in Farnborough during the war, was struggling to understand his experimental results in terms of isotopes, an interpretation totally different to the one J. J. was giving. A quick survey of the publications in scientific journals in the 1910s and 1920s, however, disproves G. P.'s statement on the scope of positive-rays research. There are only a handful of papers from other scientists following Thomson's path on positive rays.

In J. J.'s mind, Aston's developments were only 'the beginning of a harvest of results which will elucidate the process of chemical combination, and thus bridge over the most serious gap which at present exists between Physics and Chemistry' (Thomson, 1913b, preface to the second edition of 1921). He was determined to use research on positive rays as an alternative method of studying the composition and structure of the atoms by analysing the proportion and behaviour of positively charged ions – the charge of these, acquired by losing a certain number of electrons, would give information about the most likely arrangement of corpuscles in the atom – and G. P. eventually became his main supporter in this project. It was in the field of positive rays that G. P. centred his research and published his first experimental papers after the Great War (G. P. Thomson, 1920b, c, 1921, 1922).

5.5   Physics at war

As already mentioned, G. P. had early been involved in the war effort, first in the British rearguard, and from March 1915, in Farnborough, working for the Royal Aircraft Factory. By then, other Cavendish researchers like Aston and G. I. Taylor had joined the team assigned the task of improving the theoretical and practical understanding of the dynamics of air-flow: 'Aerodynamics was in an odd state in 1915' G. P. wrote in his memoirs. 'There was very little valid theory, for the classical hydrodynamics of Euler gives absurd answers, in all but a few cases, for the forces exerted by the air on bodies moving through it' (G. P. Thomson, 1966, p. 41). This work on aerodynamics resonated with at least two of G. P.'s main interests: his passion for sailing and model building, and his Cambridge training, in which mastery of the mathematical methods for dealing with the continuum was essential. His research activity in Farnborough involved first flying in planes with equipment and prototype pieces to control, and later learning how to fly. The latter was, in those conditions, a dangerous activity, since the instructors were 'not good', and 'the death rate among pupils was high … it about equalled that at the battle of Waterloo' (p. 44). His engagement with the Royal Aircraft Factory also involved a few months' work in the USA.

J. J., on the other hand, stayed mainly in Cambridge and became involved in the war effort more as an administrator than as an active researcher through his two appointments at the time: his presidency of the Royal Society and his vice-presidency of the Board for Invention and Research. These two positions gave him the opportunity to speak up on two issues on which he had strong opinions: his views on the relationship between science, the military and industry, and his ideas on education. In both areas, J. J. saw the need for much reform in Britain and he used his high-profile positions during the war to lobby for them.

Britain received a big shock during the Great War. It was apparent that the enemy forces were far ahead in terms of scientifically led and industrially developed weaponry. Classical gentlemanly fighting was over and a new kind of war was taking place, in which new, large-scale, scientifically produced weapons seemed to be determining the outcome of the war. The ability to produce new arms, both for defence and attack, depended on the joint work of scientists and the military on an industrial scale. The Royal Society, and Thomson with it, contributed to the new circumstances, since most active scientists were engaged in war-related work, thus giving 'striking evidence of the extent to which even the most recondite branches of science can find application in modern warfare' (Thomson, 1917, p. 93).

According to J. J., the war impetus had to help Britain gain momentum and make this collaboration a long-lasting one, encouraging the creation of a 'permanent establishment of both the Army and the Navy, special laboratories, properly equipped and in close touch with the services, whose work should be the discovery and development of applications of Physical, Chemical and Engineering Science for Military and Naval purposes' (Thomson, 1917, p. 94). This scheme resulted in the establishment of the Department for Scientific and Industrial Research (DSIR) in 1915 and its redesignation and enlargement once the war was over. Thomson's rhetoric advocated deeper integration of science, the military and industry, which was necessary in order to turn Britain into a powerful country, not only during the war but also in peacetime, since the power of a country needed also to be shown in the commercial arena and the provision of goods for its citizens. Britain could not base its strength on trading alone, but must also include production, since 'we have been taught by bitter experience that it is not safe to have regard to nothing but money profit in developing the industries of the country'. Relying on his own experience as the director of the Cavendish, Thomson was aware of the increasing amounts of money needed to undertake large-scale research, and the reluctance to invest in undertakings of doubtful and, certainly not immediate, economic return. That is why he proposed that the companies of one particular industrial sector should unite and create research institutions that would benefit all the companies in that sector. In his advocacy for more organized practical research, however, Thomson was concerned that industrial research should not be promoted to the detriment of pure science, since he believed in a top-down, one-way, linear relationship between pure and applied science:

> We must be careful, however, and I think this might be regarded by the Royal Society as one of its most important duties, that the badly needed increase in research in applied science is not accompanied by any slackening off in research in pure science, that is, research made without the idea of commercial application, but solely with the view of increasing our knowledge of the laws of nature. Even from the crudest utilitarian point of view, nothing could be more foolish than the neglect of pure science, for most of the great changes that have revolutionized or created great industries have come from discoveries made without any thought of their practical application (Thomson, 1917, p. 95).

This was not just opportunistic talk during the wartime emergency. In fact, J. J. repeated this discourse on many occasions, even after the war. As president of the newly created Institute of Physics, for instance, in 1923 Thomson promoted a series of lectures dedicated to the application of physics to industry. In the foreword to the publication containing these lectures, he praised the recent growth in the number of research laboratories and institutes started 'either by individual firms or, under the auspices of the Advisory Council for Scientific and Industrial Research, by combinations of firms', as a first step towards the implementation of this new strategy. The country was in a process of reconstruction after the war, and this needed 'increased production and increased opportunities of employment, such as would be afforded by the foundation of new industries ... To judge from the experience of the past no method is more likely to lead to this result than the industrial applications of physical science and research in physics' (Thomson, 1923b, foreword).

Also in that year, during his third trip to the USA, Thomson was greatly impressed by the research facilities he saw there. Invited by the Franklin Institute of Philadelphia, Thomson gave a series of lectures about which we shall speak in the next section, but he also visited a number of laboratories, such as those of General Electric and the Bell Telephone Company, where he had first-hand experience of the benefits of such large-scale scientific and industrial endeavours. In comparison with the USA, however, Britain had a better record in scientific research at the universities, and that moved J. J. to advocate a closer and symbiotic relationship between industry and academia, arguing that fundamental discoveries from the latter were beneficial to the former and, as a consequence, industry should also sponsor the work in the universities. On many occasions thereafter, he used his discovery of the electron as a clear example of the use of apparently useless academic findings in the development of new commodities:

> One of the most remarkable things about Modern Physics is that though it deals with the most recondite of phenomena, few branches have led to such important practical applications. Could anything seem at first sight less likely to be of practical utility than the electron – yet it is now the foundation of a great industry employing many men and much capital. It is the electron that makes long-distance wireless possible, and enables one human being to instruct, amuse or bore another at a range of thousands of miles [referring to the radio] … These discoveries were made without any thought of such applications, they illustrate the value of research made only for the purpose of advancing knowledge. It is discoveries made in this way that create new industries and revolutionise old ones. These applications of pure science to industries benefit not only the industries but also science itself. For when a discovery is seen to have practical applications and is taken up by the engineers it can by the resources and powerful appliance at their disposal be developed at a rate and on a scale impossible in the Laboratory, and so science advances more quickly. The Industries thus repay the debt they owe to Science, and unite with it in the task of increasing our knowledge of the wonderful universe in which we live (Thomson, 1930, p. 211).

Together with industrial research, Thomson became actively engaged in educational reforms during the war, with a particular emphasis on improving the teaching of the sciences. As a matter of fact, the Royal Society had been pushing for an increase and improvement in science education during the tenures of Thomson's predecessors as presidents, Archibald Geikie and William Crookes. A country like Britain could not take off scientifically and technologically if the most able students in the schools received minimal training in the sciences compared with their education in English, literature and the classics, including a certain mastery of Latin and Greek. If the power of the country was to be developed in its research institutions, there was a need for scientifically trained personnel, and this meant long-term education in the sciences from an early age. The main reason for encouraging the teaching of sciences in the elementary and secondary school was to prevent suitable candidates 'drifting into employment of secondary importance to the State' (Thomson, 1917, p. 96).

It can be argued that Thomson's position in this debate was not easy. On the one hand, he was by now an established Cambridge don, embedded in all the traditions of that university. But, as he had experienced in the 1880s and 1890s, during his attempts to reform the science triposes in Cambridge, an emphasis on science education to the detriment of the arts would be regarded with suspicion by some classically educated dons. That is why he had to measure carefully the arguments he used, distancing himself from the political left in his defence of scientific teaching. His 1917 address as President of the Royal Society is a good example of this kind of rhetorical delicacy:

> The need for a greater appreciation of the value of science has
> been brought into such prominence by the war that most of those
> who have advocated the claims of science in education have not
> unnaturally laid the greatest stress on the importance of science
> to the welfare, the power, and even the safety, of the nation.
> The supporters of literary studies have, on the other hand, dwelt
> mainly on the fact that literature broadens a man's horizon, and
> gives him new interests and pleasures, that it teaches him how to
> live, if not how to make a living. The result of this divergence of
> appeal has made the discussion appear ... almost like a discussion
> between Spirituality and Materialism, or between a saint and a man
> of business... I must protest against the idea that literature has a
> monopoly in the mental development of the individual. The study
> of science widens the horizon of his intellectual activities, and
> helps him to appreciate the beauty and mystery which surround
> him. It opens up avenues of constant appeal to his intellect, to his
> imagination, to his spirit of inquiry, to his love for truth. So far
> from being entirely utilitarian, it often lends romance and interest

to things which to those ignorant of science make no appeal to the intellect or imagination, but are regarded by them from an exclusively utilitarian point of view (Thomson, 1918, p. 186).

In those years, Thomson was not only making grandiose speeches like this one, but was also actively taking part in the partly failed reform of the educational system that Herbert Fisher, director of the Board of Education, attempted in 1917. In 1916, Thomson was appointed, by the Prime Minister, chairman of a committee to investigate the position of the natural sciences in the educational system, as part of the whole process being coordinated by Fisher. The Fisher Report was a milestone in the history of education in Britain: it famously included the extension of compulsory school education until the age of 14, and suggested the implementation of an educational system, both at primary and secondary levels, by which access to education depended on intellectual capacity only and not on social or economic status.

5.6   The electron in chemistry

In Chapter 2, we discussed J. J.'s early interest in merging physics and chemistry under the umbrella of the *physical sciences*, both at a conceptual and an institutional level. The latter project never eventuated, but, throughout his career, he kept an eye on the conceptual and theoretical rapprochement. His own research more often than not stepped into the territory of chemistry, especially chemical bonding in molecules (Stranges, 1982; Sinclair, 1987; Chayut, 1991). The problem of the nature of chemical bonding emerged in all his atomic models, starting with his early vortex ring model, where the bonds of affinity were related to the stable dynamical configurations of the rings. From 1891 onwards, the paradigm of electrolysis that permeated his research on discharge tubes introduced the notion of electrically polarized atoms and the possibility that chemical bonding was directly related to electrical forces. Once the electron-corpuscle came on the stage as part of the structure of all atoms, J. J. turned it also into a very promising tool to explain, at least in part, the mechanism of chemical bonding.

The appropriation of the electron by the chemists was not, however, straightforward; in 1903, during his lectures in Yale, Thomson expressed regret that, although the idea that the chemical forces were of electrical origin had had many supporters (from Davy and Faraday, to Berzelius and Helmholtz), 'chemists seem, however, to have made but little use of this idea, having apparently found the conception of "bonds of affinity" more fruitful' (Thomson, 1904a, p. 133). Twenty years later, in 1923, Thomson was back in America, delivering a series of lectures at the Franklin Institute, in Philadelphia, immediately to be published as *The Electron in Chemistry*. In between, a generation of American chemists had taken seriously much of Thomson's theoretical chemistry and developed the foundations of a theory of bonding based on the electronic atomic structure, often quoting J. J. Thomson as the originator of such an approach.[3]

From the very beginning, J. J.'s models of the atom were closely linked to the question of chemical bonding. The arrangement of the corpuscles in different layers, with a maximum in each layer, was related to the periodicity of chemical properties reflected in Avogadro's system. Atoms tended to stability, and this might explain the ease with which they gained or lost external corpuscles in the presence of other atoms, which, in turn, would enable an electrostatic bond between the oppositely charged atoms. As we saw in the previous chapter, this idea reinforced his faith in the reality of Faraday tubes, since these might be the physical counterpart of the lines used by chemists to represent the structure of chemical formulae. In his widely read 1907 book, J. J. ruled out the existence of other kinds of mechanism for chemical bonding, even though he had to leave many compounds, including most organic substances, unexplained. In Arabatzis' words, 'in the second decade of the twentieth century, because of the popularity of Thomson's theory among American chemists, the exclusive reality of polar bonds became part of the orthodoxy of the day' (Arabatzis, 2006, p. 181 see also Chayut, 1991).

But, just before the war started, and moved by the evidence from his experiments on positive rays, J. J. changed his mind and began to talk about two types of chemical bonding: one polar and the other non-polar. The argument for the latter was the following. There were quite a number of molecules, like H_2 or CO, in which, after dissociation in the cathode, he could observe parabolas of equal intensity corresponding to the two component elements on both the positive and the negative side. This symmetry should mean that, before the splitting, the two atoms in the molecule could not have opposite charges and that their bond was not polar. He had not drawn this conclusion by the time of the first edition of *Rays of Positive Electricity*, but made it the topic of a paper in *Philosophical Magazine* only in 1914.

The analogy he used in this paper to justify the existence of chemical bonding without the need for total ionization of its component atoms was the existence of physical properties like 'intrinsic pressure and surface tension of liquids, latent heat of evaporation, cohesion of solids and liquids, the rigidity of

---

[3] According to Chayut (1991, p. 541) these were: George Falk, John M. Nelson, Hal T. Beans, Harry S. Fry, Lauder W. Jones, William A. Noyes, Julius Stieglitz, William Bray, Gerald Branch, S. J. Bates, E. C Baly, C. H. Desch, G. N. Lewis, and Irving Langmuir.

solids, and so on' (Thomson, 1914, p. 757). These would be the result of forces between *neutral* molecules stemming from the organisation of the electrical charges in their interior. Analogously, one could imagine chemical bonding as the result of the organization of corpuscles inside the atom without the need to lose or gain any number of them. The key to this new type of bonding was two-fold: the tendency of all atoms towards 'saturation' and the existence of Faraday tubes. According to J. J., the former was achieved when the atom had eight corpuscles in the outmost layer. In that case, corpuscles were fixed in their positions. When their number was less than the maximum eight, corpuscles were mobile inside the atom, a mobility that was only limited by the fact that every corpuscle was linked to the positive part of the atom – which he by now accepted occupied a central position in the atom – by a Faraday tube. This allowed for the possibility of a particular corpuscle gaining stability, not by abandoning the atom, but by having its Faraday tube ending in the positive part of another nearby atom:

> Now let us consider how the corpuscles in these atoms can be fixed. They are not fixed when the atom is by itself. In this case the tube of force starting from a corpuscle in the atom, returns to a positive charge in the same atom and possesses considerable mobility, as the corpuscle at one end of it can move freely about in the atom. The corpuscle will not be fixed unless the tube of force at its end is anchored to something not in the atom, i.e. it must end on another atom. Thus if there are *n* free corpuscles in the atom, to fix these and thus saturate the atom, the *n* tubes of force which start from the *n* corpuscles must all end on other atoms and not return to the original atom. Thus to ensure saturation from every free corpuscle in an atom, a tube of force must pass out of that atom and end on some other, and this must hold for every atom in the molecule. When the atoms are electrically neutral, i.e. have no excess of positive over negative charge or *vice-versa,* for each tube of force which passes out of an atom, another must come in; and thus each atom containing *n* corpuscles must be the origin of *n* tubes going to other atoms and also the termination of *n* tubes coming from other atoms (Thomson, 1914, p. 782).

As was often the case, J. J. did not explain the mechanism by which this dislocation of the Faraday tubes might take place, but he merely emphasized the explanatory power of this model for a large number of molecules, especially organic compounds. The first example he used to illustrate this new bonding was, however, his recently 'discovered' X_3 molecule, which he had interpreted as corresponding to H_3. In this new model, this molecule was possible if each H atom was the origin of a Faraday tube going to a second H atom but receiving a tube from the third atom, creating a sort of triangular structure. The importance of this new idea resides in the fact that this new theory of J. J. Thomson of chemical bonding was one of the main catalysts for the American chemist Gilbert N. Lewis to formulate his theory of chemical bonding based on the sharing of two electrons by two atoms in 1916 (Kohler, 1971), which was generally accepted by the community of physical chemists by the time of Thomson's third visit to the USA in 1923.

Lewis trained in chemistry at Harvard at the end of the nineteenth century. These were the years of the emergence of physical chemistry with the ionic theory of Svante Arrhenius, and, like many people of his generation, he spent the usual postdoctoral year in Germany, with Ostwald and Nernst. Back in the USA, he took the position of instructor first at Harvard, and from 1905 at the MIT, and began to think, teach and speak (but not to publish) about the electron theory of chemical valence. His main image was that of a cubic atom, with the eight saturation electrons occupying the vertices of the cube. At the time, Thomson and most physicists were concerned with the stability of moving electrons in atoms due to radiation, but that was not a main issue for chemists. Their real interest was in the stability of chemical compounds, and they tended to picture electrons in fixed positions and imagine the relocation of electrons to create polar bonds.

With Thomson's legitimization of non-polar bonds from 1914, the critical voices against the exclusivity of the polar bond gained momentum, culminating in Lewis' seminal 1916 paper 'The atom and the molecule', in which he introduced the notion of 'electron-sharing' within his cubic structure of the atom. His idea was that atoms could share electrons by being close enough so as to have one or more electrons occupying a position in the external electronic layer of both atoms. The contemporary reader may be surprised that this static model was proposed at a time when Bohr had already introduced his quantum atom. Lewis mentioned it in this paper but dismissed it, together with any other planetary model, on the basis that it 'is inadequate to explain even the simplest chemical properties of the atom' (Lewis, 1916, p. 773) and that it was only an attempt, and a failed one according to Lewis, to account for electromagnetic problems of stability. Furthermore, the paper contains a kind of manifesto that resonates with Thomson's methodology:

> Indeed it seems hardly likely that much progress can be made in the solution of the difficult problems relating to chemical combination by assigning in advance definite laws of force between the positive

and negative constituents of an atom, and then on the basis of these
laws building up mechanical models of the atom. We must first of
all, from a study of chemical phenomena, learn the structure and the
arrangement of the atoms, and if we find it necessary to alter the law
of force acting between charged particles at small distance (Lewis,
1916, p. 773).

Lewis' model and its subsequent modifications became very popular among
chemists during and immediately after the Great War. The static atom (not
necessarily cubic) not only gave a reasonable explanation of chemical bonding, but also seemed to receive support from early X-ray diffraction in metals
(Stranges, 1982, chapter 8). Two post-war papers by Thomson also supported
the idea of a static atom rather than a dynamic one à la Bohr (Thomson, 1919,
1920). While physicists, especially in Europe, increasingly regarded J. J. as a
relic from old times gone by, he was regarded as a kind of foundational father
of the new chemistry of valence, especially in America. There, he had his tri-
umphant 1923 trip, visiting and lecturing at universities on the East Coast as
well as major industrial research laboratories like General Electric's facilities in
Schenectady, where he met chemists such as Irvin Langmuir and Albert Hull,
both key players in the development of Lewis' ideas.

The Electron in Chemistry is a five-chapter book that situates Thomson's career at the centre of the current status of chemistry (Thomson, 1923a). He was
father of the two main scientific developments that had finally shattered the
division between physics and chemistry: the electron and the study of positive
rays provided information about the structure of the atom and, with it, a radical change in chemists' understanding of previously purely phenomenological
concepts like affinity. If Thomson's view of chemistry was, in the 1880s, that
it was an under-age science just beginning to acquire some epistemological
maturity, this book has a celebratory flavour to it: chemistry was, at last, a
grown-up science, thanks to the input from physics, the most important of
which were *his*. This triumphalism is clear in the preface of the book:

> It has been customary to divide the study of the properties of matter
> into two sciences, physics and chemistry. In the past the distinction
> was a real one owing to our ignorance of the structures of the atom
> and the molecule. The region inside the atom or molecule was an
> unknown territory in the older physics, which had no explanation
> to offer as to why the properties of an atom of one element differed
> from those of another element. As chemistry is concerned mainly
> with these differences there was a real division between the two
> sciences.

> In the course of the last quarter of a century, however, the
> physicists have penetrated into this territory and have arrived at
> conceptions of the atom and molecule which indicate the way in
> which one kind of atom differs from another and how one atom
> unites with others to form molecules. These are just the problems
> which are dealt with by chemists and thus if the modern conception
> of the atom is correct the barrier which separated physics from
> chemistry has been removed.
>
> From many points of view the chemical side seems to be the one
> on which the most striking developments of the newer physics may
> be expected (Thomson, 1923a, preface).

Not surprisingly, J. J. used his particular pedagogical style, stressing the achievements of his atomic and molecular model, and avoiding what his static model of the atom could not explain. If, for instance, Lockyer's spectroscopic experiments had been a support for his 1897 suggestion of a subatomic particle, now, in 1923, there was no mention of the very complicated spectroscopic phenomena for which a static model of the atom could not account. Quantum theory was, of course, totally absent, as was any reference to Bohr and Sommerfeld, or even to Rutherford's 1919 demonstration of the existence of the proton. For Thomson, the main source of information about the positive part of the atom was his work on positive rays which showed that 'the carrier of the positive charge, unlike that of the negative, varies from element to element', only to concede that 'as the mass of the positive charge is always an integral multiple of a unit, it is natural to suppose that this mass is made up of a number of units bound together' (Thomson, 1923a, p. 2).

To be sure, 1923 was also the year of the publication of Lewis' book *Valence and the Structure of Atoms and Molecules*, which not only fixed the main concepts in the discussion of non-polar atomic bonding, but also introduced a first rapprochement between the Thomson–Lewis static ideas and the Bohr–Sommerfeld dynamic atom, considering the now three-dimensional orbits as static, irrespective of the actual position of the electron within the orbit. And soon afterwards, the whole new field of quantum chemistry finally set aside the fundamental tenets of J. J. and his classical approach to physics (Gavroglu & Simoes, 2000; Gavroglu, 2001). Were it not for the events we shall explore in the next chapter, the 1923 trip might well have turned into Thomson's swan-song.

6

The electron in Aberdeen: from particle to wave

6.1 Professorship in Aberdeen

On 13 June 1922, the Professor of Natural Philosophy at the University of Aberdeen, Charles Niven, retired, vacating the chair that Maxwell himself had occupied some decades earlier. Niven, together with his elder brother William, and Horace Lamb, formed 'the first generation of Cambridge Maxwellians' (Warwick, 2003a, p. 325–33). Graduating as senior wrangler in the Cambridge Mathematical Tripos in 1867, Charles Niven had made important early contributions to the understanding of Maxwell's *Treatise*, showing that 'it was the direct application of the new equations to problems in electrostatics and, especially, electrical currents that they identified as the obvious avenue of research' (p. 329), and not the dynamic foundation of the theory. But, by 1922, it had been a long time since he had done any relevant scientific research and his resignation was awaited with anticipation.

G. P. Thomson applied for this position. The testimonials he enclosed with the application letter, nine in total, all came from well-established scholars directly related to Cambridge: E. C. Pearce, Master of Corpus Christi College; William Spens, Fellow, Tutor and Director in Science, Corpus Christi College; Ernst Rutherford, director of the Cavendish Laboratory; Horace Lamb, former professor in Manchester, retired in Cambridge since 1920; Alex Wood, University Lecturer in Experimental Physics, Fellow and Tutor of Emmanuel College; R. A. Herman, University Lecturer in Mathematics, Fellow of Trinity College; B. Melvill Jones, Francis Mond Professor of Aeronautics in Cambridge; C. T. R. Wilson, Solar Physics Observatory, Cambridge; and R. T. Glazebrook, former demonstrator at the Cavendish and now Director of the National Physical Laboratory. In some of these testimonials, we can read sentences unusual in this kind of letter, such as 'I have known Mr George Thomson of Corpus Christi College since his school days, and have watched his career with much interest', or 'I have known Mr Thomson for a number of years, dating back to a period much before his undergraduate days, and have followed his career with great interest' (letters by Wood and Glazebrook, 21 July 1922, UAb). For obvious reasons, the application did not include a testimonial from his father, who, in G. P.'s words, 'has been my chief teacher' (G. P., application letter, UAb).

From the few records kept about his appointment, it is possible to infer that G. P. was the only candidate for this position: The only surviving file in the archives of the University of Aberdeen pertaining to this competition contains only G. P.'s application letter together with the testimonial letters on his behalf; and the Minutes of the Aberdeen University Court speak only of one candidate. Furthermore, the retiring professor's brother, the late Sir William D. Niven, had been G. P.'s godfather, which makes it plausible that G. P. was particularly encouraged to apply for this job due to the close relationship between the Niven family and the Thomson family (Thomson, 1936, p. 43).

In any case, G. P. Thomson was appointed professor on 5 September 1922, and was granted the sum of £1,600 to equip the totally out-of-date laboratory with new instruments. With the appointment and the money for research, it appeared only natural to him to continue the research he had been doing in Cambridge. As a consequence, he turned his research laboratory into a virtual extension of his father's basement room at the old Cavendish. He bought the tubes and the vacuum pumps he needed, and reproduced in Aberdeen his father's experimental set-up in order to continue research on positive rays. G. P.'s laboratory notes show, however, that this transfer of skills from Cambridge to Aberdeen was no straightforward business. Maintaining a steady high vacuum proved to be more difficult than expected, his teaching and administrative duties were time-consuming, and the lack of research students to help with the practicalities of research was a clear limitation. At the beginning, the department consisted of him, two lecturers (Dr Fyvie and A. E. M. Geddes), a laboratory assistant (C. G. Fraser) and an administrative assistant (Miss Jack). With these conditions, he managed to publish his first experimental results only in 1926, four years after taking up his new position (G. P. Thomson, 1926a, b). By then, and with a grant from the Department of Scientific and Industrial Research that allowed him to have a personal assistant to help run experiments, he managed to consolidate a small research department in Aberdeen, one that was virtually an extension of his father's old Cavendish.

Like his father, G. P. Thomson met his future wife in the context of the university. On the day of his job interview, he was invited to dine by the Principal and

Vice-Chancellor of the university, Sir George Adam Smith, who 'was undoubtedly the best loved man in the University' (G. P. Thomson, 1966, p. 70). At dinner, 'I sat next to [his daughter] Kathleen and was much impressed', and 'before I had been long in Aberdeen I had made the first of many proposals', until 'on the evening of 19 July [1924] she got tired of refusing me and for a change herself proposed' (p. 72). The wedding took place on 18 September in the chapel of King's College, in Aberdeen, and the new family settled in a beautiful house, 'rather too large for us and somewhat costly', about five miles from town. G. P. would normally go by bus to the university and only from time to time in his own car. The family had four children (John, Clare, David, and Rose).

Academically, the Cavendish connection, as it were, had not disappeared with his transfer to Aberdeen. G. P. frequently went down to Cambridge to discuss ideas and, during the summer vacation, even to perform some experiments in the Cavendish Laboratory. For instance, in a paper of 1926, he acknowledged advice and help received from C. D. Ellis and Mr. Wooster 'for their kindness in helping me to use the microphotometer in the Cavendish Laboratory ... and to Sir Ernest Rutherford for permission to use it' (G. P. Thomson, 1926c, p. 421). Further evidence in support of this Cavendish connection comes from the letters of his wife Kathleen, which help us trace his relatively frequent trips to Cambridge (G. P. Archives, A14 A). A third kind of source of support are his presentations at the Kapitza Club on February and August 1927, and March 1928 and July 1929 (Churchill College Archives, CKFT 7/1).

A glance, however cursory, at the major features of G. P.'s initial experimental set-up is essential in order to understand what happened in the period 1926–1928. Figure 6.1. shows the schematic representation of G. P.'s experiments. A cathode-ray tube on the right with the cathode perforated is the source of positive rays. These are introduced into the tube on the left, where they can be deflected by the electric and magnetic fields, scattered by other gases, or made to interfere with matter in some other way. A photographic plate on the extreme left of the tube records the impact of the positive rays after their interaction with the fields or matter. The experiments he had started in 1920 in the Cavendish, and which he was continuing in Aberdeen, consisted of a study of the scattering of these positive rays by different elements and molecules. G. P. and J. J. supposed this to be a valid way to study the binding force between atoms in molecules. However, even the only two papers he managed to produce on this topic while in Aberdeen show major uncertainties in regard to the validity of the experimental results and their analysis. One of the major issues was the impossibility of discriminating between atomic and molecular positive rays in the incident beam. Furthermore, the charge of the positive rays was not necessarily constant all the way through the scattering chamber, since they

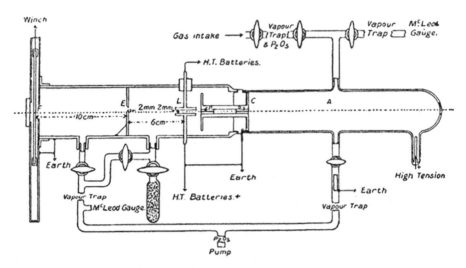

Figure 6.1. Experimental set-up for the experiments on positive rays. A cathode ray tube (right) sets the positive ions of the gas in motion. These are deflected in L and photographed in F. (G. P. Thomson, 1920c, p. 241). Courtesy: Taylor and Francis, Ltd.

could easily gain and lose electrons. Last but not least, G. P. was working and analysing his results on the Newtonian assumption that the scattering between atoms and molecules was governed by an inverse-square law, as with macroscopic bodies, an assumption that did not fit his experimental results at all.

While he was operating valves, sealing glass tubes, and fine-tuning the vacuum pump, G. P. was also trying to keep up-to-date with theoretical developments in physics. Thus, he was well aware of de Broglie's principle through a paper recently translated into English probably by Fowler. In that paper, de Broglie presented the results from his recent Ph.D. dissertation, on the basis of which he was 'inclined to admit that any moving body may be accompanied by a wave and that it is impossible to disjoin motion of body and propagation of wave' (de Broglie, 1924, p. 450). This is what soon came to be understood as the principle of wave–particle duality, and which Schrödinger eventually turned into a full formulation of a wave quantum mechanics (Raman & Forman, 1969).

De Broglie's paper was entitled in English 'A tentative theory of light quanta', a title which had very strong resonances for the Thomson family. The nature of light and other radiations had been a topic of heated debates for 20 years, a debate in which J. J. had been one of the main proponents. Actually, in the same year – 1924 – J. J. Thomson was working on his nth attempt to explain the photoelectric effect and the nature of light within his metaphysical framework in which the ether, and the Faraday tubes in it, were essential elements. Far

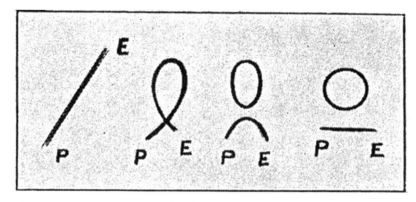

Figure 6.2. A Faraday tube between the electron and the proton bends and forms a photon. (Thomson, 1925, p. 24). Courtesy: Cambridge University Press.

from denying the experimental evidence for the quantum of light, J. J. stressed that this quantification was only the result of a process in the continuous medium. Figures 6.2 and 6.3 show very graphically his idea for the processes of light emission and absorption, respectively, in the simple case of a hydrogen atom. Assuming, as he did, that the proton (P) and electron (E) in the atom interacted by means of a Faraday tube connecting them, one could imagine what happened to the tube when an electron 'jumped' from one orbit of high energy to an orbit of lesser energy. The Faraday tube would first bend, and then form a loop that would detach from the original tube: this would constitute the emission of a photon. Similarly, a quantum of radiation, in the form of a closed-loop Faraday tube, could be absorbed by the tube uniting a proton and an electron, providing the energy for the electron to jump to a higher energy state (Thomson, 1924, 1925).

This theory could be relegated to a cabinet of intellectual curiosities, were it not for the impact it had on his son. The paper by de Broglie was an attempt to formulate a new theory of light, as much as J. J.'s was. Both were published in the same year and G. P. tried to unite them in a paper in the *Philosophical Magazine*. In retrospect, G. P. would regret publishing this paper, calling it 'an example of a thoroughly bad theoretical paper' (G. P. Thomson, 1961, p. 821), even though it was proof, in his reconstructions of history, that he had paid attention to de Broglie's theory as soon as it was published in the British milieu: 'I think in retrospect I was in advance of my time, I think I paid more attention to de Broglie than probably anybody else in this country on the whole. Some people thought it was just nonsense' (G. P. Thomson, oral interview, AHQP, Tape T2, side 2, 8). There are two points to emphasize here. First, that G. P. was among the first British physicists to pay serious attention

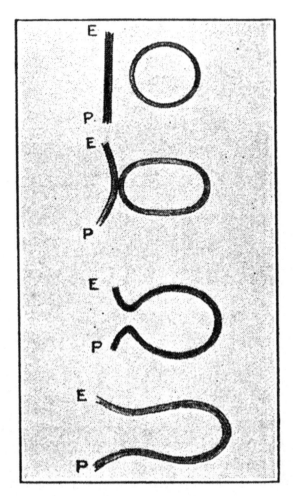

Figure 6.3. A photon is absorbed by a Faraday tube between the electron and the proton. (Thomson, 1925, p. 27). Courtesy: Cambridge University Press.

to de Broglie's theory. And second, that he understood it as a theory of light and electronic orbits, not as a theory of electron diffraction or even as a theory of free electrons.[1] Although not many, a few scientists, including de Broglie

---

[1] In his reconstruction of the events, G. P. presented a slightly different version of the facts: 'At that time we were all thinking of the possible ways of reconciling the apparently irreconcilable. One of these ways was supposing light to be perhaps particles after all, but particles which somehow masqueraded as waves; but no one could give any clear idea as to why this was done. The first suggestion I ever heard which did not stress most of all the behaviour of the radiation came from the younger Bragg, Sir William Lawrence Bragg, who once said to me that he thought the electron was not so simple as it looked, but never followed up this idea. However, it made a considerable impression on me, and it pre-disposed me to appreciate de Broglie's first paper in the Philosophical Magazine of 1924' (G. P. Thomson, 1961, p. 821).

and Einstein, as well as M. Born, J. Frank, and W. Elsasser in Göttingen and, possibly, also C. D. Ellis in Cambridge were, by 1925, already thinking about de Broglie's principle in terms of electron diffraction (Russo, 1981, pp. 141–4). As we shall see, the idea of electron diffraction as an experimental application of de Broglie's theory came to G. P. Thomson only some time in the summer of 1926, not in 1924.

The title of G. P.'s 1925 paper is 'A physical interpretation of Bohr's stationary states', and in it he tried to explain de Broglie's radical hypothesis in terms of J. J.'s metaphysical framework. If the trajectories of the electrons were understood in terms of waves, only those orbits whose circumference is a multiple of the wavelength can constitute stable orbits around the nucleus, a suggestion that was totally in tune with Bohr's quantification. G. P.'s suggestion was that these stationary states could be equally achieved in his father's 1924 atomic model explained above. If the proton and the electron were united by a Faraday tube of force, 'it will thus be able to transmit waves, and the condition that will be taken as determining the possible states is that the vibrations in this tube shall be in tune with the period of the orbit' (G. P. Thomson, 1925, p. 163). Thus, the physical properties of the Faraday tube uniting the electron with the nucleus would determine the waves accompanying the movement of the electron and, therefore, the possible orbits in an atom.

This example shows once again the son's intellectual continuity with his father's ideas. Faraday tubes of force were very much a Thomson working tool, and G. P. was maintaining the tradition. We can see this not only in the 1925 paper but also in the notes of 1923 for his lectures in Aberdeen. The lectures on electricity to the Senior Honours Class contain the following definition, which reflects G. P.'s use of Faraday tubes: 'Def^n. Regarded as filling all space. Originally an attempt to interpret the phenomena of electricity as residing in the medium. Now perhaps best regarded as a mathematical dodge but may have physical reality in a modified form' (G. P. Archives, F4, 7). It is worth noting that G. P.'s attachment to this concept played a paradoxical role. While being an increasingly problematic and out-of-fashion theoretical tool, it was his attachment to Faraday tubes that predisposed him to pay particular attention to de Broglie's work well before most British physicists did. At the same time, as the last sentence of his lecture notes suggests, tubes of force were, for G. P., a heuristic tool with perhaps less physical reality than they had for his father. The science of the son, while highly dependent on the father's, was not a simple reproduction of it, but an evolution from it. And this continuity could be better maintained in Aberdeen, where G. P. Thomson had virtually extended the dominions of the J. J.'s old physics into a new environment. Experimentally, the devices he developed in his new laboratory turned it into an extension of his father's rooms at the Cavendish. Intellectually, no-one in Aberdeen was going to react against this physics as the young generation of 'modernists with a vengeance' did in Cambridge (Hughes, 1998).

6.2   Electron diffraction

'By 1926 I was feeling depressed by having failed to produce anything of real note. In fact, positive rays, as distinct from the study of isotopes, were nearly worked out, at least for the time' (G. P. Archives, A6, 7). Looking at his laboratory notebooks, however, no hint of G. P.'s disappointment is evident: in the first half of the year he kept accumulating data and changing the experimental conditions in his work on positive rays. The last entry before the summer break was from 23 June, he was testing the scattering of positive rays in argon; the next entry, on 23 August, clearly signalled a shift of research project: '*Alteration to apparatus*. A slip of gold leaf mounted on brass carrier + partly covering aperture in camera. Aperture in camera enlarged downwards about ¼. Diaphragm slit moved down to zero mark' (G. P. Archives, C24, 13, emphasis in the original). His quest for electron diffraction had started.

The various autobiographical notes by G. P. on the events leading up to his measurement of electron diffraction are a little incomplete and not wholly consistent with each other in some details. They all agree, however, as does all the other evidence, in attributing critical importance to the month of August 1926, both in Oxford and in Cambridge. From the 4th to the 11th, the British Association for the Advancement of Science held its annual meeting in Oxford; and it turned out to be the forum in which many British and American physicists learnt about the latest developments in wave quantum mechanics. During the spring of that year Erwin Schrödinger had, on the basis of de Broglie's ideas, reinterpreted wave mechanics from a quantum perspective. Max Born, present at the meeting, explained these developments to the participants, and the topic became one of the highlights in the informal discussions at the meeting. Born's paper had a strong impact on many present, but especially on Clinton J. Davisson, an American physicist working at the Bell laboratories, when he heard that the anomalous results he had been obtaining in experiments on electron dispersion with his colleague Lester H. Germer might be signs of electron diffraction. That part of the story, which was studied in detail by Arturo Russo, ended with the confirmation of electron diffraction in the Bell laboratories and the sharing of the Nobel Prize with G. P. Thomson for their experimental proof of de Broglie's principle. At the time of his first experiments, however, Thomson was not fully aware of Davisson's project. Born also mentioned the experiments of the young German physicist Walter M. Elsasser, who had unsuccessfully tried to detect diffraction patterns in the transmission of an electron beam through a metallic film (Russo, 1981).

Straight after the Oxford meeting, G. P. stopped over in Cambridge, where he usually spent part of the summer with his family. There, he could continue discussions on electron diffraction, and, at the Cavendish, he met E. G. Dymond who thought he had obtained evidence of electron diffraction with experiments on the scattering of electrons in helium (Dymond, 1926; G. P. Thomson, 1968). It is the case that, be it from conversations in Oxford or in Cambridge,[2] G. P. saw – or was led to understand – that his experimental device in Aberdeen was almost all that was needed to try electron diffraction through solids. Actually, the role played in this story by Dymond's results, later to be proved erroneous, is quite unclear. In one of the accounts, G. P. said that Dymond's experiments were slightly demoralising, since the 'only' thing left for him was to obtain evidence for the same phenomenon in solids: 'When I returned to Aberdeen I thought, "Well, it has apparently been done with a gas, but let's try it on solids"' (G. P. Thomson, 1961, p. 823). In other accounts, however, G. P. emphasizes that it was the uncertainty of Dymond's results that encouraged him to attempt electron diffraction through solids (G. P. Thomson, 1968, p. 7).

A quick comparison of the experimental set-up he had so far used for his experiments on positive rays (Figure 6.1) with the one he used in his work of 1926–1927 (Figure 6.4) clearly shows that few changes were needed for

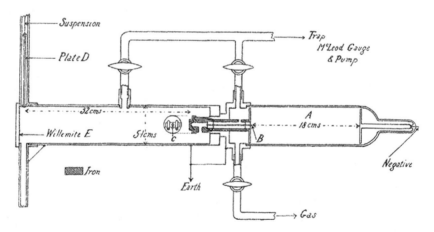

Figure 6.4. Experimental device used by G. P. Thomson in his observations of electron diffraction. A cathode-ray tube (right) sends electrons to be photographed at E after being diffracted at C. (G. P. Thomson, 1928a, p. 602). Courtesy: Royal Society.

[2] In Cambridge, P. M. S. Blackett had also tried to obtain evidence of electron diffraction, but gave up after a few months (Nye, 2004, p. 46).

the new measurements. His original set-up provided positive rays using a cathode-ray tube; and now, the same tube could be the source of a beam of electrons. The 'apparatus for studying the scattering of positive rays ... could be used for this experiment with little more change than reversing the current in the gaseous discharge which formed the rays' (G. P. Archives, A6, 10/3). The rest of the arrangement only differed in the fact that instead of scattering the electrons in a gas, he attempted their diffractive dispersion through a thin metallic plate. The latter was, in retrospect, the only real experimental change, one for which he depended on the good skills of the head mechanic in the laboratory, C. G. Fraser, who had been trained as a watchmaker, and succeeded in obtaining the extremely thin metallic films that were needed.

In fact, things proved to be a bit more complicated. According to his laboratory notebook, G. P.'s experiments between 23 August and 15 September 1926, mainly using celluloid films, did not provide much information. The next entry is dated 10 December 1926, when he decided to start afresh and set up his experimental device in the 'end room in Basement', taking much care at every step. For instance, 'scrupulous care observed in glasswork, all tubes washed before blowing + cleaned before mounting with a saturated solution of potassium bichromatic in commercial sulphuric acid afterwards washed + then dried with alcohol' (G. P. Archives, C24, p. 21), or 'high tension apparatus mounted in the corner of the room away from galvanometer + lead to discharge tube slung from shellacked glass rods hooked to cords along the roof' (p. 22). It is worth pointing out that the absence of laboratory entries in October and November is possibly due to a lack of time and space to perform his experimental work during term time, which makes us wonder how pressing he thought these experiments were in the autumn of 1926.

The December work ended up with a disappointing 'Apparatus Dismantled' on 3 January 1927, due to constant leakage in different parts of the set-up, and a new attempt to rebuild the apparatus began on 12 January. For half a year, he and his research student Andrew Reid systematically sent cathode rays through thin celluloid films, looking for possible diffraction of the electrons using photographic plates and a photometer. The first significant results were published in a joint paper in *Nature* in June 1927, where they showed one picture 'recalling the haloes formed by mist round the sun' (G. P. Thomson & Reid, 1927, p. 890). Actually, the only way to see the concentric circles produced by the sought-after diffraction of cathode rays through celluloid was to measure the intensity of light on the photographic plate with a photometer: 'in this way rings can be detected which may not be obvious to direct inspection'. These results, however encouraging, were certainly not conclusive, but they were presented as confirmation of Dymond's earlier results.

After the untimely death of Reid in a motorcycle accident, G. P. Thomson continued the experiments on his own, now no longer with celluloid, but with thin films of gold, aluminium, platinum and other metals prepared by Mr. Fraser. Again we read, in the summer of 1927, and after more than a month's pause, 'New Apparatus, July 7th 1927' (G. P. Archives, C24, p. 71), with which he immediately started testing the new films. We often read sentences like the following, which give us a sense of the mechanical difficulties of dealing with such thin films, once the pumps, cathode rays and other more sophisticated elements of the experimental set-up were finally under control: 'Films mounted but broke in the drying' or 'Several films produced + floated on water but not successfully lifted due to extreme difficulty of seeing whether the film lay across the aperture of the holder or not. Films left overnight floating on water were found to have thinned down or else disappeared with the action of the residual acid in the water' (p. 84). In October 1927, he obtained some good photographs indisputably showing the diffraction patterns of cathode rays through some of these thin films as he had been expecting for over a year now. These results were published in a note in *Nature* in December 1927, followed by a full account of his work in several articles in the *Proceedings of the Royal Society* in 1928 and 1929 (G. P. Thomson, 1927, 1928a, b, 1929a). Electron diffraction was now a reality.

Before we move on, another element needs to be highlighted here: the close connection between his experiments and the long tradition in research on X-ray diffraction, to which G. P. was certainly no stranger. After the discovery of X-ray diffraction by Planck's protégé, Max von Laue, in Munich in 1912, G. P.'s lifelong friend Willie Bragg had modified his father's research project on X-rays and understood that X-ray diffraction could be used as a tool to determine the crystalline structure of solids. This other father–son story culminated in the shared Nobel Prize that the two Braggs received in 1915 and, most importantly, consolidated the emergence of the new science of X-ray crystallography in Britain.[3] G. P. certainly followed these developments closely through his friendship with the young Bragg, with whom he spent summer holidays on his boat, the *Fortuna* (Hunter, 2004, pp. 70 and 104). His other lifelong friend, C. G. Darwin, was responsible for the formulation of the most successful theory of X-ray diffraction between 1913 and 1922 (G. P. Thomson, 1963).

There were quite a few similarities between G. P.'s experiments and the ones on X-ray diffraction, since the energy of the waves de Broglie was talking about

---

[3] I do not mean to say here that there is a parallel between the story of the Braggs, father and son, and the story of the Thomsons, also father and son. I only suggest that one can easily suppose that G. P.'s friendship with Bragg was a natural channel for him to follow closely the developments on X-rays.

was of the same order as that of hard X-rays. The only great difference between X-rays and the waves of cathode rays was that the latter could be deflected with electric and magnetic fields due to their electric charge, a difference that proved essential in order to confirm that the diffracted patterns were due not to secondary X-rays but to the cathode rays themselves. Again, this was a feature that the experimental arrangement of G. P.'s project on positive rays already included: like the experiment that had led to the hypothesis of the corpuscle in 1897, the Thomsons' study of positive rays involved their deflection by electric and magnetic fields in the glass tube.

The following anecdote serves to illustrate the importance of electromagnetic deflection. Probably around the beginning of March 1928, G. P. had the opportunity to discuss his experimental results with Schrödinger himself as the latter recalled in 1945:

> After mentioning briefly the new theoretical ideas that came up in 1925/26, I wish to tell of my meeting you in Cambridge in 1927/28 (I think it was in 1928) and of the great impression the marvellous first interference photographs made on me, which you kindly brought to Mr Birthwistle's [sic] house, where I was confined with a ... cold. I remember particularly a fit of scepticism on my side ('And how do you know it is not the interference pattern of some secondary X-rays?') which you immediately met by a magnificent plate, showing the whole pattern turned aside by a magnetic field (Schrödinger to G. P. Thomson, 5 February 1945, G. P. Archives, J105, 4).

The pictures G. P. obtained were powerful enough to convince his audience (Figure 6.5). The circular halos were widely recognized as the Hull–Debye–Scherrer patterns of diffraction, already known for X-rays. Therefore, if those pictures were really obtained from cathode rays, there was no choice but to accept that the electrons behaved like waves: 'The detailed agreement shown in these experiments with the de Broglie theory must, I think, be regarded as strong evidence in its favour' (G. P. Thomson, 1928a, p. 608).

If the period between the summer of 1926 and the spring of 1928 saw only a few changes in the experimental culture of G. P. Thomson, there is more to be said about the impact that his experimental results had on the theoretical foundations of his physics. The photographs were, for him, a clear demonstration that the electrons behaved like waves, and solid evidence of the validity of de Broglie's principle, a principle that involved a serious reconsideration of the nature of waves and particles. In a paper he submitted in November 1927 paper, we can read that his results involved 'accepting the view that ordinary Newtonian mechanics (including the modifications for relativity) are only a

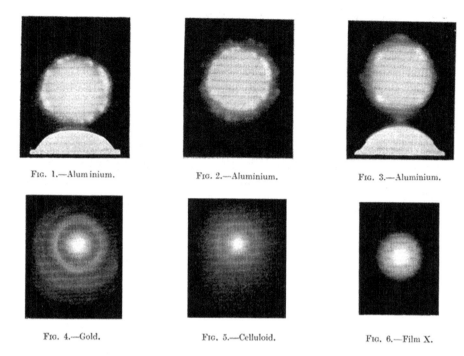

FIG. 1.—Aluminium.   FIG. 2.—Aluminium.   FIG. 3.—Aluminium.

FIG. 4.—Gold.   FIG. 5.—Celluloid.   FIG. 6.—Film X.

Figure 6.5. First patters of electron diffraction obtained by G. P. Thomson. (G. P. Thomson, 1928a, plate 19). Courtesy: The Royal Society.

first approximation to the truth, bearing the same relation to the complete theory that geometrical optics does to the wave theory' (G. P. Thomson, 1928a, pp. 608–9). This and other similar sentences appear to be suggesting a partial abandonment of the classical mechanics he had thus far been immersed in, immediately raising the question of what it was that G. P. had understood and accepted of the latest developments in quantum physics.

Besides the impetus that the British Association for the Advancement of Science meeting in Oxford of 1926 gave to a significant number of British physicists, G. P. benefitted, once again, from his close friendship with C. G. Darwin who in 1924 became the Tait Professor of Natural Philosophy in Edinburgh and who spent the spring of 1927 in Copenhagen, where he could discuss the latest developments in quantum mechanics with Bohr, Heisenberg, and Schrödinger, among others. On his way back, Darwin spent some time in Aberdeen, in G. P.'s home. In this way, G. P. learnt the new wave mechanics from Darwin's explanations: 'we had long talks about all this, and really began to get an idea about it' (G. P. Thomson, oral interview, AHQP, Tape T2, side 2, 15). The timing was just right. As G. P. was seeing with his own eyes the diffraction patterns of cathode rays, he understood them in light of Darwin's explanations. There were obvious reasons, on G. P.'s side, to think that Darwin was possibly the British physicist best suited for understanding them at that time. In his biographical memoir on Darwin, G. P. said that 'I am inclined to think that his most useful work was as an interpreter of the new quantum theory to experimental physicists.... I should like to record my great debt to him for the many ideas in physics he helped me to understand' (G. P. Thomson, 1963, p. 81).

Later, we shall explore the impact that the experiments on electron diffraction had on G. P.'s theoretical framework, looking at his early explanations of the phenomena, and also at the relationship between his and Darwin's explanations. However, let us pause for a moment before that and look at the reaction of his father, J. J., in the face of the inescapable experimental evidence.

6.3   The father's interpretation

The father saw, in the experiments of his son, the final proof of his lifelong metaphysical project and a clear sign of the invalidity of quantum physics as an *ultimate* explanation. His world had always been, and still was, a world of ether, in which discrete entities, including the electrons, were but epiphenomena in the ether. Now, in 1928, J. J. Thomson felt his metaphysical idea had proved true and that electron diffraction was a sign that discrete models of matter were only rough approximations of reality. In his mind, the 'very interesting theory of wave dynamics put forward by L. de Broglie', and experimentally demonstrated by his son, was not in contradiction to classical mechanics. In the first of a series of papers he published in *Philosophical Magazine*, J. J. tried to show that 'the waves are also a consequence of classical dynamics if that be combined with the view that an electric charge is not to be regarded as a point without structure, but as an assemblage of lines of force starting from the charge and stretching out into space' (Thomson, 1928a, p. 191).

Thomson had never accepted the idea put forward by Larmor and Lorentz at the turn of the century that an electron was a point charge of electricity in the ether. Now, the detection of a train of waves associated with the motion of electrons was, for him, proof that he had been right: Maxwell's equations did not predict such a wave for a point-electron, and *therefore* such a view of the electron had to be wrong. On the other hand, de Broglie's wave could be obtained on purely classical grounds if he assumed the electron to be a two-part system: a 'nucleus which ... is a charge e of negative electricity concentrated in a small sphere' (Thomson, 1928b, p. 1259), and a sphere surrounding it 'made up of parts which can be set in motion by electric forces ... consist[ing] either of a distribution of discrete lines of force, or of a number of positively- and negatively-electrified particles distributed through the sphere of the electron'

(p. 1254). Using this ad hoc structure, J. J. deduced the relationship between the speed of an electron and the wavelength of its sphere to be the same as that expected by de Broglie and measured by G. P. As for the electron as a composite sphere, he would eventually express this in terms of what he came to call 'granules', particles 'having the same mass μ, moving with the velocity of light c, and possessing the same energy μc^2' (Thomson, 1930/31, p. 86).

At a conference held at Girton College, Cambridge, in March 1928 entitled 'Beyond the Electron', J. J. argued that talking about a structure for the electron was not ludicrous. Thirty years earlier, when he first suggested that corpuscles would be constituents of all atoms, thus initiating the exploration of the structure of the atom, he was accused of alchemy. The developments of the physics of the electron had dismissed that accusation. Now he felt justified in talking about the structure of the electron in the light of the latest developments by his son. 'Is not going beyond the electron really going too far, ought one not draw the line somewhere?', he would ask rhetorically, to which he would reply that 'It is the charm of Physics that there are no hard and fast boundaries, that each discovery is not a terminus but an avenue leading to country as yet unexplored, and that however long the science may exist there will still be an abundance of unsolved problems and no danger of unemployment for physicists' (Thomson, 1928c, p. 9).

The diffraction experiments showed that 'we have energy located at the electron itself, but moving along with it and guiding it, we have also a system of waves' (Thomson, 1928c, p. 22). Following the similarities with his structure of light of 1924, he supposed that the electron 'had a dual structure, one part of this structure, that where the energy is located, being built up with a number of lines of electric force, while the other part is a train of waves in resonance with the electron and which determine the path along which it travels' (p. 23). For him, the association of a wave with an electron was not a new phenomenon. The association had already been made when, in the late eighteenth century, the corpuscles of light that Newton had postulated needed to be complemented by wave explanations. It was not so strange to see that the new corpuscles, the electrons, had to receive similar treatment. Furthermore, discussions on the nature of light in the previous two decades had paved the way for the acceptance of the duality of the electron.

In the world of J. J, electron diffraction brought about the possibility of challenging, rather than accepting, the new quantum physics. A continuous metaphysics in which all phenomena and entities could be seen as structures of the ether was, in his view, still possible. Furthermore, J. J. felt that at last electron diffraction provided the clinching argument in defence of the old worldview, something that the developments of the previous two decades had, albeit seemingly, jeopardized. Electron diffraction was proof of the complexity of the electron and, therefore, of the validity of classical mechanics. Quantification of magnitudes such as momentum or energy 'is the result and expression of the structure of the electron; only such motions are possible, or at any rate stable, as are in resonance with the vibrations of the underworld of the electron' (Thomson, 1928c, p. 31).

At the root of his models, there was a metaphysical problem as much as an epistemological one. As already stated, J. J.'s metaphysics involved a continuum in terms of which all discrete phenomena could, and should, be explained. Parallel to that was an epistemological problem: for Thomson, de Broglie's theories, as much as Planck's, were valid only from a mathematical point of view. Their results were valid, but they did not entail real, true physics. And that was the strength that J. J. saw his theory had over de Broglie's: 'The coincidences are remarkable because two theories could hardly be more different in their points of view. M. de Broglie's theory is purely analytical in form; the one I have brought before you ... is essentially physical' (Thomson, 1928c, p. 34). It comes as no surprise that, true to the spirit in which he was educated and with which he had worked for so long, physical meant mechanical.

In an ironic remark on the situation of physics in previous years, he stated in 1930 that

> when the waves are taken into account, the classical theory of dynamics
> gives the requisite distribution of orbit [of the electrons] in the
> atom, and as far as these go the properties of the atom are not more
> inconsistent with classical dynamics than are the properties of organ
> pipes and violin strings, in which, as in the case of the electron, waves
> have to be accommodated within a certain distance. It is too much to
> expect even from classical dynamics that it should give the right result
> when supplied with the wrong material (Thomson, 1930, pp. 26–7).

Obviously, the fact that the proof had come from within the family could not but be an added reason to rejoice.[4] Actually, we see him using the expression *your* electronic waves' in writing letters to his son after G. P.'s experiments (G. P. Thomson, 1964, p. 157). In the decade between these events and his death in 1940, J. J. did not change his mind. The last paper he ever published, sent in October 1938 at the age of 81, still proclaimed his son's experiments as proof of the validity of the old classical mechanics (Thomson, 1939).

---

[4] Oral interview with G. P. Thomson, *Archive for the History of Quantum Physics*, Tape T2, side 2, 9: 'Well, I think he was very pleased [with my developments], largely because it was in the family'.

Let us finish this section with the manuscript notes of a lecture 'to discuss a method of describing the electric field in terms of physical conceptions rather than mathematical symbols' (Thomson, CUL, Add 7654, UC7.7, p. 1) dated 1932–3. True to the philosophy of science he emphasized in the 1890s, already described in Chapter 3, J. J. kept arguing for the superiority of what he calls 'the physical method', which consists of visualizing every physical event through some mental process. He complained that 'of late years the tendency has been to adopt a purely mathematical treatment of some branches of physics, to make no attempt to visualize what is going on but to regard everything from the point of view of the properties of a function which satisfies a certain differential equation', in clear reference to the dominant trend in quantum physics, a trend that, in his view, would paralyse science altogether, since 'you cannot visualise an abstraction, while you can a physical system, and visualization is a tremendous aid to concentration of thought and to the birth of new ideas'. Of course, mathematical precision is of great value in scientific work, but certain so-called precision is not really such, and can also be a science-stopper: mathematicians usually obtain only the differential equations, and not their full solutions; they speak about wave-functions of electrons, but they do not know how to interpret the function. They have, 'generally speaking, succeeded in obtaining solutions when the bodies they are dealing with have simple geometrical shapes, when they are for example planes, cones or spheres, … [It] resembles one in a picture by a cubic artist built up of figures just mentioned with an ellipsoid or two in by way of ornament' (p. 2–3). Using an old expression of his, he reminds us that he regards 'a physical theory as a tool and not as a creed … the important thing about a theory is not so much is it "true" as it is useful' (p. 3). In that same lecture, he defended what he called the 'materialist' attitude towards elusive entities like the ether, and he used two examples to support his thesis: the existence of atoms and the fluid theory of electricity. The first was a usual response against the claims of positivists, but the latter comes a bit as a surprise:

> When the fluid theory was first introduced physicists took a comparatively materialistic view of these fluids. They were prepared to consider the possibility of their having some of the ordinary properties of fluids, mass for example, and experiments were made to see if the absence or presence of these fluids affected the weight of a body. By the middle of the last century the materialistic view had gone quite out of fashion, the word fluid was interpreted in a very picturesque sense, it was a *disembodied* fluid from which everything material had disappeared and it was thought almost as improper to suppose it had

any material attributes as to suppose an *angel* possessed any limbs other than wings. The fluid of those times was like the grin of the Cheshire cat without the cat. The progress of science has justified the more materialistic view. It has shown that negative electricity consists of a collection of electrons, that these electrons have all the same mass and charge, and that a free charge of negative electricity may be regarded as a gas made up of discrete particles ... The electric fluid is an example of the very interesting fact in the history of science that to tone down the material side of a theory to escape ... criticism ... has generally resulted in growing up the substance to get the shadow (Thomson, CUL, Add 7654, UC7.7, p. 7).

As we shall see in the next section, the battle in the late 1920s and early 1930s between those who advocated a realist interpretation of the new quantum mechanics, especially in its wave formulation, and those defending a more mathematically oriented and less intuitive approach, especially in the formalism of Heisenberg and Dirac, was being won by the latter. J. J. Thomson's understanding of what physics *had to be* was no different from that of many people of his generation and younger. Electronic waves could be the beginning of their comeback. Electronic waves were proof that the ether was real and, to a large extent, material.

6.4   The son's reaction

G. P. made public his preliminary results in a short note in *Nature* dated May 1927 and in a presentation at the Kapitza Club, in Cambridge, on the 2 August (G. P. Thomson & Reid, 1927; Churchill Archives, CKFT, 7/1). In November, he was ready to publish a long and detailed paper in the *Proceedings of the Royal Society*, preceded by another short note in *Nature* (G. P. Thomson, 1927; 1928a). These papers were basically a just description of the experimental methods and results from which he extracted what were, for him, two obvious consequences: the results could only be explained if we considered electron beams to undergo diffraction, like X-rays, and this diffraction was consistent with the one predicted by de Broglie. In other words, G. P. Thomson had no doubt that his experiments were a direct proof of the principle of wave-particle duality; but, as for any further consequences of this principle, he preferred to be cautious. A quotation from his Friday speech at the Royal Institution of 6 June 1929, describes his approach to what we may call metaphysical speculation in this period of his life. After explaining in full detail the experiments on electron diffraction, he ventured to try to answer the 'great difficulties of interpretation. What are these waves? Are they another name for the electron itself? … Some of these questions I should like very briefly to discuss, but we now leave the sure foothold of experiment for the dangerous but fascinating paths traced by the mathematicians among the quicksands of metaphysics' (G. P. Thomson, 1928d, p. 281). Despite this reluctance, there were two questions that he was wont to address in these metaphysical excursions: the reality of the ether, and the ontological status of the electronic wave in relation to the particle.

The best and most comprehensive source we have for understanding G. P.'s views at this time is *The Wave Mechanics of Free Electrons* (G. P. Thomson, 1930), which contains the lectures he gave at Cornell University during the last term of 1929. Here we find a thorough explanation of the implications of his experiments on the very important question of the existence of the ether. The wavelengths of electron waves and X-rays are in the same range, but electron waves and X-rays clearly behave differently, for the first can be deflected, and the second cannot. If that is the case, one might need to assume two different media to account for the different behaviour of the two waves, 'but it is not a very attractive idea to have two ethers filling the space, especially as the waves of protons – if they exist – would demand yet a third. Space is becoming overcrowded' (G. P. Thomson, 1930, p. 11). G. P.'s solution was to apply Ockham's razor, doing away with the ether and retaining the wave formulation; 'perhaps simple physicists may be content as long as the waves do their job guiding the electron, and it is possible that, after all, the question will ultimately be seen to be meaningless' (p. 12). In the lecture at the Royal Institution mentioned above, he stated that 'The easiest way of looking at the whole thing seems to be to regard the waves as an expression of the laws of motion' (G. P. Thomson, 1928d, p. 282). And to lend authority to his point of view, he finished his speculations by quoting Newton's famous 'hypotheses non fingo'.

Abandoning the implications of an ether was, however, not straightforward. One of G. P.'s most radical, and short-lived, speculations during these years was the possibility that strict energy conservation might have to be abandoned in order to explain beta radioactive decay (Jensen, 2000). That comes as no surprise since C. G. Darwin had been, for almost a decade, a strong advocate of this possibility (Stolzenburg, 1984, pp. 13–9, 67–9, 317–9; Kalkar, 1985, pp. 6, 91–9, 305–19, 347–9). Actually, it was following calculations made by Darwin on his way back from Copenhagen that G. P. suggested a *mechanism* to account for the dispersion of energy in beta decay. Essentially, G. P. suggested that the actual beta emission did conserve energy, but that the huge acceleration experienced by the electron in its ejection from the nucleus involved the creation of an energetic wave, like 'the sound produced by the firing of an atomic gun whose bullet is the electron' (G. P. Thomson, 1929b, p. 410). Such a wave could be supposed to 'possess energy when highly concentrated which it loses on spreading out' (p. 415), giving rise to an indeterminacy in the energy of the electron. This idea, however, involves some friction of the pulse wave, which is an implicit remnant of the role of the ether (G. P. Thomson, 1928c). G. P. did not pursue this idea any further, but this instance shows us both that the abandonment of classical concepts was not straightforward, and that his incursion into the quantum physics was very strongly dependent on Darwin's understanding of the new physics (Navarro, 2009).

In Darwin's view, Schrödinger's wave mechanics had some sort of ontological priority over matrix mechanics. It is part of the received view on the developments of quantum physics that in 1926 the equivalence between wave and matrix mechanics was demonstrated (Muller, 1997; Kragh, 1999; Mehra & Rechenberg, 2000). However, Darwin's work in these years shows his uneasiness with Heisenberg's methods, since they do not provide true, i.e., mechanical, explanations. In several places, we find statements like the following: 'There are probably readers who will share the present writer's feeling that the methods of non-commutative algebra are harder to follow, and certainly much more difficult to invent, than are operations of types long familiar to analysis'. And as a solution to this difficulty, he thought that 'Wherever it is possible to do so, it is surely better to present the theory in a mathematical form that dates from the time of Laplace and Legendre, if only because the details of the calculus have been so much more thoroughly explored' (Darwin, 1928, p. 654; see also Darwin, 1931, p. 124). Furthermore, in what looks like very clear philosophical positioning, both epistemologically and metaphysically, Darwin argues for the metaphysical reality of the wave function:

> We shall … take the wave function ψ as the central feature of the quantum theory … From the practical point of view the great advantage of thinking in terms of ψ is that it forces on our attention the diffractive effects of matter and treats them as a more fundamental property than the ray-like properties which suffice for the description of ordinary events… There is no need to invoke particle-like properties in the unobserved parts of any occurrence, since the wave function ψ will give all the necessary effects (Darwin, 1929, pp. 391–2).

According to Darwin, matrix mechanics is only an ingenious mathematical method useful for explaining observable results, especially in the field of spectroscopy, but it is not helpful in understanding what the reality of things is. Darwin's understanding of quantum physics emphasized the link between a continuous wave-like metaphysics and the discrete quantum manifestations in natural phenomena. The latter was only particular instances of a much richer world, the world of possibilities.

G. P. Thomson's interpretation was, although slightly different, related to Darwin's. He regarded his experiments as some sort of proof of de Broglie's principle, and this meant, for him, that matter could altogether be thought of as essentially continuous:

> Matter is still supposed made of discrete units, but instead of these units moving according to laws which concern them alone, as did the laws of Newtonian dynamics, we have had to introduce laws based on waves. Now a wave is essentially a continuous thing, even if the continuity is only mathematical. It is spread through space, not divided into little lumps. *So although the older belief in the discontinuity of matter still holds, it has lost some of its rigidity; continuity has crept in by the back door* (G. P. Thomson, 1930, p. 12).

Quantum physics was, in this way, no longer a threat to the continuous conceptions of matter with which he had grown up. Certainly, continuity was no longer dependent on the physical existence of the ether, but the fundamental entities of nature, electrons and quanta, were proving to be *also* fundamentally continuous. In this respect, Darwin's influence was instrumental to G. P.'s understanding the new physics as, in some way, being in continuity with the old. Darwin emphasized the natural link between both approaches in a very revealing statement: 'We recall that Hamilton worked out very exact analogies between the behaviour of rays of light and particles of matter … But de Broglie pushed the matter to its logical conclusion by saying that if light and matter are refracted in the same way, then they ought to be diffracted in the same way too' (G. P. Thomson, 1930, p. 107). From Darwin's point of view, the new physics in its wave formulation was a natural extension of classical physics and could preserve a continuous ontology. This approach enabled G. P. to embrace electron diffraction as a radical experimental result, and wave–particle duality as equally a revolutionary principle, without totally undermining his fundamental understanding of physics, both mathematically and metaphysically. Certainly, both Darwin and Thomson could feel at ease with a quantum mechanics that preserved a continuous ontology, as well as the need for visual explanations, since these were fundamental tenets in their training as physicists. As for visual interpretations, an example G. P. often used in his popular lectures is that of the gossamer spider:

> When at rest this spider is a minute insect. When it wants to move it sends out streamers into the air, and floats away owing to the action

of the air on these filaments which stretch out a foot or more all round it. Just so the electron, when it is part of an atom its waves are limited to that atom, or even to a part of it. They are curled round on themselves, as it were. Suppose, now, an electron escapes from the hot filament of a wireless valve and gets free. Its waves will spread far out into the space round it. I regard it as still a particle at the centre of its wave system. The analogy can be pressed further. If the wind sweeps the spider past an obstacle the filaments will catch. The pull on filaments will move the spider, and he will feel that there is something in the way, even though his body does not actually hit it. In the same way the waves are a means by which the motion of the electron is affected by things which the main body of the electron never comes very near (G. P. Thomson, 1929c, p. 220).

This analogy resonates with J. J.'s suggestion of a structure for the electron; and it can also be seen as a pedagogical explanation of Darwin's idea that the wave function describes all the possible movements of the electron. The three approaches are certainly not totally equivalent, but they are linked by the rejection of an ultimate exclusively discrete, i.e., quantum, physics. The three were aware that the diffraction experiments entailed a turning point in physics, albeit a turning point that allowed for continuous explanations of Nature to regain their legitimacy against the threat of an excessively discrete quantum physics.

6.5   Moving to London. Electron diffraction turns into an instrument

The Aberdeen experience came to an end in 1930, when G. P. was offered a chair at Imperial College, London, after his close friend Willie Bragg had declined the same offer. Also in that year, he was made a fellow of the Royal Society, following in the steps of both his father and his (maternal) grandfather. The new appointment was an opportunity to work in a larger setting than Aberdeen, with more means of conducting research, although it also involved undertaking more administrative work than in Scotland. His work on electron diffraction had made him a well-known figure in contemporary physics in his own right, not as just the 'son of Sir J. J.': he had obtained proof of one of the most unintuitive principles of quantum mechanics, 'the son of Sir J. J.', having obtained proof of one of the most unintuitive principles of the new quantum mechanics. However, he was no theoretician, and his experience in physics was basically on the experimental side. And, actually, beyond the popular talks in which he discussed the nature of electronic waves, he was well aware that the deep theoretical implications of his discovery were beyond his research interests and capabilities. Thus, his experiments were not a springboard for him to become part of the quantum generation, but rather he capitalized on his experience with electron diffraction as an experimentalist, by turning his discovery into a laboratory instrument. In his Nobel lecture, he gave a *biography* of the electron, from his father's discovery to his own experiments, but then cautioned his audience, saying that electron diffraction was not 'of interest only to those concerned with the fundamentals of physics. It has important practical applications to the study of surface effects' (G. P. Thomson, 1938). While the Braggs had turned X-ray diffraction into a technique for the investigation of the structure of crystals, electron waves could complement the X-ray technique for thin surfaces, since X-rays are too penetrating for the analysis of just a few layers of atoms, while it is precisely the fast absorption of electron waves that make their diffraction useful for thin layers of solid metals. This became G. P.'s project in the 1930s. His last paper from Aberdeen, and the articles published in the early 1930s, when he was already at Imperial College, relate to what he called 'for shortness an electron camera' (G. P. Thomson & Fraser, 1930, p. 641). G. P. wanted the device, for which he provided a detailed technical configuration in the *Proceedings of the Royal Society*, to be an easily reproducible apparatus 'to study the diffraction patterns formed by the reflection of cathode rays from crystalline surfaces'.

After his experiments, and those of Davisson and Germer, were made public in 1927, a relatively large number of research teams managed to obtain similar results: C. Eckart and F. Zwicky at Caltech, A. L. Patterson in Berlin, E. Rupp at AEG, Berlin, D. C. Rose in Cambridge, S. Kikuchi and S. Nishikawa in Tokyo, A. Szczeniewski in Paris, R. T. Cox, C. G. McIlwraith and B. Kurrelmeyer in New York, A. F. Joffé and A. N. Arsenieva in Leningrad, among others (Gehrenbeck, 1974, pp. 309–34). Most of these replications followed G. P. in the use of higher-energy electrons, as opposed to Davisson and Germer's use of the more difficult-to-handle slow rays, but they also followed the latter in the study of beams incident and reflected on the surfaces of metallic solids rather than the beams traversing the very thin foils that G. P. had used. This 'electron camera' put these two elements together, creating a tool that merged the new electron diffraction with the more settled techniques of X-ray crystallography. That also explains why electron diffraction quickly became a tool, since it could be appropriated by crystallographers with relative ease.

The undulatory nature of the electron was soon accepted, both by quantum theoreticians who, in a way, had been expecting the result, and by experimentalists, who immediately made use of electronic waves. After G. P.'s and

Davisson and Germer's experiments, there was no question of the existence of these waves. There were, of course, many question marks over their behaviour in particular instances, and this is where we find controversies: in refraction and dispersion, polarization of electrons, some peculiar results with mica, ruled gratings, surface gases, and gaseous scattering, for instance (Gehrenbeck, 1974). And electron diffraction became a tool for examining anomalous results and open questions obtained using the techniques of X-ray diffraction. The following extract from a coauthored paper by researchers at Imperial College, including G. P. Thomson, bears witness to the naturalness with which this transfer took place:

> In May, 1930, Mr. H. M. D. White at our request carried out at University College an X-ray examination by the Debye–Scherrer method of two platinum films which had been sputtered for this purpose on the quartz rods. The results were, however, negative. Accordingly, in March, 1931, it was suggested to Professor G. P. Thomson, FRS, that an electron diffraction examination of these platinum surfaces might possibly disclose that feature of the structure upon which catalytic properties appeared to depend (Finch *et al.*, 1933, p. 415).

In January 1936, G. P. had to postpone all his commitments due to serious illness; a perinephric abscess with subsequent complications for which he had to undergo surgery five times during 1936 and 1937. He spent part of his convalescence in Cambridge, in J. J.'s Trinity home, rather than in London. Due to this long illness, 'I was in bed when I heard that C. J. Davisson and I had been jointly awarded a Nobel Prize [of 1937] for the discovery of electron diffraction. This was an excellent tonic, but it was decided that it would be unwise to go for the formalities of the presentations on December 10th' (G. P. Thomson, 1966, p. 88). He was not able to give his Nobel speech until the spring of 1938, when he made a special trip to Stockholm. Interestingly, G. P. did no further work on electron diffraction, but moved to the study of slow neutrons, artificial radioactivity, and, in a matter of months, the dangerously promising field of artificial disintegration of nuclei and atomic energy. The last would eventually make him the chairman of the MAUD committee in charge of exploring the possibilities of Britain developing an atomic weapon during the World War II.

6.6   End of an epoch

On 3 February 1933, amidst the struggles of the Depression years, the Cavendish Laboratory formally opened a new building for low-temperature and magnetic research under the leadership of Pyotr Kapitza. Lord Rutherford, the head of the laboratory for almost 15 years, had managed to obtain funds from the Royal Society and the Department of Scientific and Industrial Research to expand the research capability in fundamental physics and to consolidate the laboratory as a world centre of experimental physics. The Cavendish was savouring the fruits of success, after the so-called *annus mirabilis* of 1932 (Hughes, 2000). James Chadwick's identification of the neutron, John Cockcroft and John Walton's nuclear disintegration, and Patrick Blackett's experiments on positrons, gave the laboratory unprecedented fame within and outside physics. The influential journalist J. G. Crowther, a good friend of the laboratory, wrote in early 1934 that its 'reputation had become magnificent beyond the hope of sustainment but mortal expectation has been disproved by the last thirteen years of its history' (Crowther, 1934, p. 7). *The Times* reported nationwide on the opening of the Mond Laboratory, emphasising that Kapitza's work was 'a striking example of the cooperation of several scientific bodies', and that scientific research was moving to a scale that 'cannot be supported by the relatively small income of an institution like the Cavendish Laboratory' (*The Times*, 3 February 1933). The following day, *The Times* also underlined the importance of pure research for the benefit and the welfare of the nation: 'If careful planning can help scientific research, then the Mond Laboratory should increase our knowledge of nature as notably as its neighbour, the Cavendish, has done; and out of that new knowledge it is quite possible that new industries will arise to bring comforts or further sources of power to man' (4 February 1933). And in a more popular way, the report in the *Manchester Guardian*, probably written by Crowther, said: 'Let the sceptics be assured. No doubt the tangible benefits of this beautiful laboratory … should ultimately filter down to the buying public in the shape of better dance music' (*The Manchester Guardian*, 3 February 1933).

In his speech for the opening ceremony, the Chancellor of the University and by then Lord President of the Council, Mr Stanley Baldwin, said:

> The development of industry to-day is depending more and more
> on the application of new ideas and discoveries in pure science, and
> successful industrial research is ultimately dependent on the vigorous
> prosecution of research in pure science with the object of adding to
> our knowledge of the processes of nature and generally without regard
> to any practical application. Experience has shown that many of the
> most important applications of science to industry have resulted from
> such fundamental researches (*The Times*, 4 February 1933).

The rhetoric of this speech emphasized the fact that pure research was not an intellectual game of interest only to academic scientists. The welfare of the people and the progress of society depended on pure research, since many new commodities were the direct result of discoveries in fundamental science. Thus, both industry and the state should finance the establishment of new scientific facilities such as the Mond Laboratory. As we saw in the previous chapter, the Great War had greatly changed the perception of scientific education and research, making the progress of science part and parcel of national development and security, and Rutherford, the new director of the Cavendish, managed to capitalize on the new mood to obtain large-scale funding for research. The school of radioactivists he built up in Cambridge during the 1920s was partly a result of the new funding possibilities. He pushed for large-scale investment of public funds in pure science, with which he completely changed the economic organisation of the Cavendish.

Around the time of the building of the Mond laboratory, old J. J. was still influential at the Cavendish, and took part in the great campaign to justify investment in apparently useless pure science. In several public lectures, he depicted his own research as one that embodied the importance of pure science for the social and economic development of the country. The discovery of the tiniest particle in the world would have proved a mere curiosity were it not, according to his line of argument, for the fact that it had spawned a large industry, spanning domestic electricity and radio broadcasting that was now creating a lot of jobs, promoting social progress:

> One of the most remarkable things about modern physics is that though it deals with the most recondite of phenomena, few branches have led to such important practical applications. Could anything seem at first sight less likely to be of practical utility than the electron – yet it is now the foundation of a great industry employing many men and much capital. It is the electron that makes long-distance wireless possible, and enables one human being to instruct, amuse or bore another at a range of thousands of miles [in reference to his own broadcasting]. The application of X-rays and radium to medicine and surgery has proved supremely important for saving life and diminishing suffering, ... These discoveries were made without any thought of such applications, they illustrate the value of research made solely for the purpose of advancing knowledge. It is discoveries made in this way that create new industries and revolutionize old ones (Thomson, 1930, p. 211).

We can discern the reference to industrial and economic output more explicitly in a recorded instructional documentary of October 1934, where he says:

It is hardly an exaggeration to say that each scientific discovery contains the germ of a new industry. Scientific discoveries are very efficient means to creating employment. And instead of attempting to reduce unemployment by reducing research, as some have suggested, I think the best hope for a durable cure is to get the hair of the dog that beat you and to go in for more and more research. In my opinion it is in laboratories and not only in the Houses of Parliament that the cure will be found (Institution of Electric Engineering, video recording 1934).

Besides the obvious political and economic agenda of these and other statements, J. J. remained, to the last days of his scientific life, attached to *his* electron and the waves it carried as sign of a possible inner structure. News was coming from the Cavendish and elsewhere of new particles and new theories, but none of these would have been possible without the electron. He liked to be portrayed as the father of a generation that could listen to the radio, use domestic electricity, and enjoy the many benefits of modern life, much of which came from practical applications of his discovery of the electron. Furthermore, he wanted to be seen as the father (or one of the fathers) of modern physics, and not a member of the last generation of what some began to refer to as classical physics. The discovery of the electron was a major step in the history of physics, akin to the breakthroughs of people like Galileo, Newton, Dalton and Maxwell. In the popular lecture broadcast on the BBC in January 1930 already referred to, he stated that 'the electron is the keystone of Modern Physics, and direct research on its properties one of the most important fields of research'. Furthermore, and to emphasize the continuity of the physics of the 1930s with the previous tradition, he stressed that 'the recent discovery of electronic waves, and the modification in the theories which it involves, is an indication that these are still fluctuating, and that one who writes about the tendencies of modern physics is liable to find his views out of date almost before they can be published' (Thomson, 1930, p. 211). J. J. thought the last battle had not yet been fought, one that would preserve the old physics, since new knowledge and new discoveries would prove the continuity between the new and the old.

His last years were quiet and peaceful in the Master's Lodge of Trinity College, fulfilling his duties at the college, which basically involved just social events. He now found time to look after the garden (botany having been a passion of his since his childhood) and to write his memoirs, which were published just before his eightieth birthday, in 1936. Thereafter, his memory and his mental abilities began to decline. J. J. was the last in a long series of Masters of Trinity College to remain in office for life, a centuries-old tradition that changed with his successor, George Macaulay Trevelyan. Thus, he stayed in the Master's Lodge of the college until his death on 30 August 1940. His ashes were soon afterwards buried in Westminster Abbey, close to the memorials of Newton, Herschel, Kelvin, and Rutherford.

References

Arabatzis, T. (2006). *Representing Electrons. A Biographical Approach to Theoretical Entities.* Chicago: Chicago University Press.

Buchwald, J. Z. (1988). *From Maxwell to Microphysics: Aspects of Electromagnetic Theory in the last quarter of the Nineteenth Century.* Chicago: Chicago University Press.

Buchwald, J. Z. and Warwick, A. eds. (2001). *Histories of the Electron. The Birth of Microphysics.* Cambridge, MA: The MIT Press.

Campbell, N. (1909). The study of discontinuous phenomena. *Proceedings of the Cambridge Philosophical Society,* **15**, 117–36.

Campbell, N. (1910). Discontinuities in light emission. *Proceedings of the Cambridge Philosophical Society,* **15**, 513–25.

Cardwell, D. (2003). *The Development of Science and Technology in Nineteenth-Century Britain.* Aldershot: Ashgate.

Cavendish Laboratory (1910). *A History of the Cavendish Laboratory 1871–1910.* London: Longman, Greens & Co.

Chayut, M. (1991). J. J. Thomson: the discovery of the electron and the chemists. *Annals of Science,* **48**, 527–44.

Crowther, J. G. (1934). *The Progress of Science.* London: K. Paul, Trench, Trubner & Co. Ltd.

Crowther, J. G. (1974). *The Cavendish Laboratory 1874–1974.* New York: History Science Publications.

Dampier, W. C. D. (2004). Liveing, George Downing (1827–1924), revised by Frank A. J. L. James, *Oxford Dictionary of National Biography,* Oxford: Oxford University Press. Available at http://www.oxforddnb.com/view/article/34559.

Darrigol, O. (1994). The electron theories of Larmor and Lorentz: a comparative study. *Historical Studies in the Physical Sciences,* **24**, 265–336.

Darrigol, O. (2000). *Electrodynamics from Ampère to Einstein.* Oxford: Oxford University Press.

Darwin, C. G. (1928). The wave equations of the electron. *Proceedings of the Royal Society,* **118**, 654–80.

Darwin, C. G. (1929). Collision problem in wave mechanics. *Proceedings of the Royal Society*, **124**, 375–94.

Darwin, C. G. (1931). *The New Conceptions of Matter*, London: Bell and Sons, Ltd.

Daston, L. ed. (2000). *Biographies of Scientific Objects*, Chicago: Chicago University Press.

Davis, E. A. and Falconer, I. (1997). *J. J. Thomson and the Discovery of the Electron* London: Taylor & Francis.

de Broglie, L. (1924). A tentative theory of light quanta. *Philosophical Magazine*, **47**, 446–58.

de Solla Price, D. J. (1957). Sir J. J. Thomson, O. M., F. R. S. *Novo Cimento Supplement*, **4**, 1609–29.

Dolby, R. G. A. (1976). The case of physical chemistry. In *Perspectives on the Emergence of Scientific Disciplines*, eds. G. Lemand *et al.*, pp. 63–74. Chicago: Aldine.

Drude, P. (1900a). Zur Elektronentheorie der Metalle. *Annalen der Physik*, **306**, 566–613.

Drude, P. (1900b), Zur Elektronentheorie der Metalle. *Annalen der Physik*, **308**, 369–402.

Dymond, E. G. (1926). Scattering of electrons in helium. *Nature*, **118**, 336–37.

Engels, F. (1845/1887). *The Condition of the Working Class in England in 1844*. New York: John W. Lovell Co. Available at http://www.marxists.org/archive/marx/works/1845/condition-working-class/ch04.htm.

Falconer, I. (1985). *Theory and Experiment in J. J. Thomson's work on Gaseous Discharge*. Unpublished Ph.D. Thesis.

Falconer, I. (1987). Corpuscles, electrons and cathode rays: J. J. Thomson and the 'Discovery of the Electron'. *British Journal for the History of Science*, **20**, 241–76.

Falconer, I. (1988). J. J. Thomson's work on positive rays, *Historical Studies in the Physical Sciences*, **18**, 265–310.

Falconer, I. (1989). J. J. Thomson and 'Cavendish physics'. In *The development of the Laboratory: Essays on the Place of Experiment in Industrial Civilization*, ed. F. James, pp. 104–17. London: Palgrave Macmillan.

Finch, G. I., Murison, C. A., Stuart, N. and Thomson, G. P. (1933). The catalytic properties and structure of metal films. Part I. Sputtered Platinum. *Proceedings of the Royal Society*, **141**, 414–34.

FitzGerald, G. F. (1897). Dissociation of atoms. *The Electrician*, **39**, 103–4.

Fraucher, L. (1844). *Manchester in 1844: Its Present Condition and Future Prospects*. London: Simpkin, Marshall and Co.

Gauld, A. (1968). *The Founders of Psychical Research*. London: Rouledge & K. Paul.

Gavroglu, K. (2001). The physicists' electron and its appropriation by the chemists. In Buchwald & Warwick (2001), pp. 363–400.

Gavroglu, K. and Simoes, A. (2000). Quantum chemistry in Great Britain: developing a mathematical framework for quantum chemistry. *Studies in the History and Philosophy of Modern Physics*, **31**, 511–48.

Gehrenbeck, R. K. (1974). *C. J. Davisson, L. H. Germer, and the Discovery of Electron Diffraction*. Unpublished PhD, University of Minnesota.

Gibson, C. R. (1911). *The Autobiography of an Electron*. London: Seeley.

Gooday, G. (1990). Precision measurement and the genesis of physics teaching. *British Journal for the History of Science*, **23**, 25–51.

Haley, C. (2002). *Boltheads and Crucibles. A Brief History of the 1702 Chair of Chemistry at Cambridge*. Cambridge: Cambridge University Press.

Harman, P. M. (1982). *Energy, Force and Matter. Conceptual Development of Nineteenth-century Physics*. Cambridge: Cambridge University Press.

Harman, P. M. (1995). *The Scientific Letters and Papers of James Clerk Maxwell*, vol. 2. Cambridge: Cambridge University Press.

Haynes, R. (1982). *The Society for Psychical Research, 1882-1982. A History*. London: Macdonald.

Heilbron, J. L. (1981). *Historical Studies in the Theory of Atomic Structure*. New York: Arno Press.

Heimann, P. (1972). The Unseen Universe. Physics and the philosophy of nature in Victorian Britain. *British Journal for the History of Science*, **6**, 73-9.

Hertz, H. (1883). Versuche über die Glimmentladung. *Annalen der Physik und der Chemie*, **19**, 782-816.

Hilken, T. J. N. (1967). *Engineering at Cambridge University, 1783-1965*. Cambridge: Cambridge University Press.

Hudson, R. (1989). James Jeans and radiation theory. *Studies in the History and Philosophy of Science*, **20**, 55-77.

Hughes, J. (1998). 'Modernists with a vengeance'. Changing cultures of theory in nuclear science, 1920-1930. *Studies in the History and Philosophy of Modern Physics*, **29**, 339-67.

Hughes, J. (2000). 1932: The annus mirabilis of nuclear physics? *Physics World*, **13**, 43-8.

Hughes, J. (2003). Occultism and the atom. The curious story of atoms. *Physics World*, November, 31-5.

Hughes J. (2005). Redefining the context: Oxford and the wider world of British Physics, 1900-1940. In *Physics in Oxford 1839-1939*, eds. R. Fox and G. Gooday, pp. 267-300. Oxford: Oxford University Press.

Hughes, J. (2009). Making isotopes matter: Francis Aston and the mass-spectrograph. *Dynamis*, **29**, 131-165.

Hunt, B. (1991). *The Maxwellians*. Ithaca: Cornell University Press.

Hunter, G. K. (2004). *Light is a Messenger: The Life and Science of William Lawrence Bragg*. Oxford: Oxford University Press.

Innes, P. D. (1907). On the velocity of the cathode particles emitted by various metals under the influence of Röntgen rays, and its bearing on the theory of atomic disintegration. *Proceedings of the Royal Society*, **79**, 442-62.

Jammer, M. (1961). *Concepts of Mass: In Classical and Modern Physics*. Cambridge, MA: Harvard University Press.

Jeans, J. (1904). *The Dynamical Theory of Gases*. Cambridge: Cambridge University Press. (second edition 1916).

Jeans, J. (1914). *Report on Radiation and the Quantum-Theory*. London: The Electrician.

Jensen, C. (2000). *Controversy and Consensus: Nuclear Beta Decay, 1911-1934*, Basel: Birkhäuser Verlag.

Kalkar, J. ed. (1985). *Niels Bohr Collected Vorks*, vol. 6. Amsterdam: North Holland.

Kargon, R. H. (1979). *Science in Victorian Manchester. Enterprise and Expertise*. Baltimore and London: The Johns Hopkins University Press.

Kim, D. W. (2002). *Leadership and Creativity. A History of the Cavendish Laboratory.* Dordrecht and London: Kluwer Academic.

Klein, M. (1973). Mechanical explanation at the end of the nineteenth century. *Centaurus,* **17**, 58–82.

Kohler, R. (1971). The origin of G. N. Lewis's theory of the shared pair bond. *Historical Studies in the Physical Sciences,* **3**, 343–76.

Kragh, H. (1999). *Quantum Generations: A History of Physics in the Twentieth Century.* Princeton: Princeton University Press.

Kragh, H. (2002). The vortex atom. A Victorian theory of everything. *Centaurus,* **44**, 32–114.

Ladenburg, E. (1907). Über Anfangsgeschwindigkeit und Menge der photoelektrischen Elektronen in ihrem Zusammenhange mit der Wellenlänge des auslösenden Lichtes. *Deutsche Physikalische Gesellschaft,* **9**, 333–48.

Larmor, J. (1900). *Aether and Matter.* Cambridge: Cambridge University Press.

Larsen, E. (1962). *The Cavendish Laboratory, Nursery of Genius.* London: Edmund Ward Ltd.

Lewis, G. N. (1916). The atom and the molecule. *Journal of the American Chemical Society,* **38**, 762–85.

Lewis, G. N. (1923). *Valence and the Structure of Atoms and Molecules.* Michigan: The Chemical Catalog Company.

Liveing, G. (1883). Presidetnial Address, Section B. In *Report of the British Association for the Advancement of Science, Southampton 1882,* pp. 479–86. London: J. Murray.

Liveing, G. (1885). *Chemical Equilibrium, the Result of the Dissipation of Energy.* London: George Bell.

Lockyer, N. (1874). Bakerian Lecture. Researches on spectral analysis in relations with the spectrum of the Sun. *Philosophical Transactions,* **164**, 479–94.

Lodge, O. (1914). *Presidential Address to the British Association for the Advancement of Science, 1913.* New York and London: G. P. Putnam's Sons.

Maxwell, J. C. (1873). *A Treatise on Electricity and Magnetism.* 2 vols. Oxford: Clarendon Press. (third edition 1891).

Maxwell, J. C. (1874). Molecules. *Report of the British Association for the Advancement of Science, Bradford 1873.* In Niven (1890), pp. 361–77.

Maxwell, J. C. (1875). Atom. *Encyclopaedia Britannica,* ninth edition. In Niven (1890), pp. 445–84.

McCormmach, R. (1967). J. J. Thomson and the structure of light. *British Journal for the History of Science,* **3**, 362–87.

McCormmach, R. (1970). H. A. Lorentz and the electromagnetic view of nature. *Isis,* **61**, 459–97.

Mehra, J. and Rechenberg, H. (2000). *The Historical Development of Quantum Theory, vol. 6 Part 1.* New York: Springer.

Meinel, C. (2009). Molecules and croquet balls. In *Models: The Third Dimension of Science,* eds. S. de Chadarevian and N. Hopwood, pp. 242–75. Stanford: Stanford University Press.

Millikan, R. A. (1917). *The Electron: Its Isolation and Measurement and the Determination of some of its Properties.* Chicago: Chicago University Press.

Milne, E. A. (1952). *Sir James Jeans. A Biography*. Cambridge: Cambridge University Press.

Moon, P. B. (1977). George Paget Thomson. *Biographical Memoirs of the Fellows of the Royal Society*, **23**, 265–310.

Muller, F. A. (1997). The equivalence myth of Quantum Mechanics. Part I. *Studies in History and Philosophy of Modern Physics*, **28**, 35–61.

Myers, G. (1989). Nineteenth-century popularizations of thermodynamics and the rhetoric of social prophecy. In *Energy and Entropy. Science and Culture in Victorian Britain*, ed. P. Brantlinger, pp. 303–34. Bloomington: Indiana University Press.

Navarro, J. (2005). J. J. Thomson on the nature of matter: corpuscles and the continuum. *Centaurus*, **47**, 259–82.

Navarro, J. (2009). "A dedicated missionary". Charles Galton Darwin and the new quantum mechanics in Britain. *Studies in the History and Philosophy of Modern Physics*, **40**, 316–26.

Navarro, J. (2012). Teaching quantum physics in Cambridge: George Birtwistle's two books on quantum physics. In *Research and Pedagogy. A History of Quantum Physics through its Early Textbooks*, eds. M. Badino and J. Navarro. Berlin: Open Access Editions.

Newton, I. (1730). *Optiks*. fourth edition. St Paul, London: William Innys.

Niven, W. D. (1890). *The Scientific Papers of James Clerk Maxwell*. 2 vols. Cambridge: Cambridge University Press.

Noakes, R. (2005). Ethers, religion and politics in late-Victorian physics: beyond the Wynne thesis. *History of Science*, **43**, 415–55.

Noakes, R. (2008). The 'world of the infinitely little': connecting physical and psychical realities circa 1900. *Studies in the History and Philosophy of Science*, **39**, 323–33.

Nye, M. J. (1972). *Molecular Reality: A Perspective on the Scientific Work of Jean Perrin*. London: Macdonald.

Nye, M. J. (1993). *From Chemical Philosophy to Theoretical Physics. Dynamics of Matter and Dynamics of Disciplines, 1800–1950*. Berkeley: University of California Press.

Nye, M. J. (1996). *Before Big Science. The Pursuit of Modern Chemistry and Physics, 1800–1940*. Cambridge, MA: Harvard University Press.

Nye, M. J. (2001). Remodelling a Classic: the electron in organics chemistry, 1900–1940. In Buchwald & Warwick (2001), pp. 339–62.

Nye, M. J. (2004). *Blackett, Physics, War, and Politics in the Twentieth Century*, Cambridge, MA: Harvard University Press.

Oppenheim, J. (1985). *The Other World: Spiritualism and Psychical Research in England, 1850–1914*. Cambridge: Cambridge University Press.

Pais, A. (1986). *Inward Bound. Of Matter and Forces in the Physical World*. Oxford: Clarendon Press.

Patisson Muir, M. M. (1884). *Treatise on the Principles of Chemistry*. Cambridge: Cambridge University Press.

Planck, M. (1906). *Vorlesungen über die Theorie der Wärmestrahlung*. Leipzig: Barth.

Raman, V. V. and Forman, P. (1969). Why was it Schrödinger who developed de Broglie's ideas? *Historical Studies in the Physical Sciences*, **1**, 291–314.

Lord Rayleigh, (1942). *The Life of Sir J. J. Thomson, sometime Master of Trinity College, Cambridge*. Cambridge: Cambridge University Press.

Roberts, G. K. (1989). The liberally-educated chemist: Chemistry in the Cambridge Natural Science Tripos, 1851–1914. *Historical Studies in the Physical Sciences*, **11**, 157–83.

Ross, H. M. (2004). Dewar, Sir James (1842–1923), revised by Trevor I. Williams, *Oxford Dictionary of National Biography*, Oxford: Oxford University Press. Available at http://www.oxforddnb.com/view/article/32804.

Russo, A. (1981). Fundamental research at Bell Laboratories: The discovery of electron diffraction. *Historical Studies in the Physical Sciences*, **12**, 117–60.

Schaffer, S. (1992). Late Victorian metrology and its instrumentation. In *Invisible Connections. Instruments, Institutions and Science*, eds. R. Bud and S. E. Cozzens, pp. 23–56. Bellingham: SPIE.

Schuster, A. (1911). *The Progress of Physics during 33 years (1875–1908): Four Lectures delivered to the University of Calcutta during March 1908*. Cambridge: Cambridge University Press.

Servos, J. W. (1990). *Physical Chemistry from Ostwald to Pauling : The Making of a Science in America*. Princeton: Princeton University Press.

Sinclair, S. B. (1987). J. J. Thomson and the chemical atom. From ether vortex to atomic decay. *Ambix*, **34**, 89–116.

Smith, C. (1998). *The Science of Energy. A Cultural History of Energy Physics in Victorian Britain*. London: The Athlone Press.

Smith, C. and Wise, N. (1989). *Energy and Empire. A Biographical Study of Lord Kelvin*. Cambridge: Cambridge University Press.

Stewart, B. and Tait, P. G. (1875). *The Unseen Universe or Physical Speculations about a Future State*. London: Macmillan & Co.

Stokes, G. G. (1896). On the nature of Röntgen rays. *Proceedings of the Cambridge Philosophical Society*, **9**, 215–16.

Stolzenburg, S. ed. (1984). *Niels Bohr Collected Works*, vol. 5. Amsterdam: North Holland.

Stranges, A. N. (1982). *Electrons and Valence: Development of the Theory, 1900–1925*. College Station: Texas A&M University Press.

Stuewer, R. (1975). *The Compton Effect: Turning Point in Physics*. London: Science History Publishers.

Sviedrys, R. (1976). The rise of physics laboratories in Britain. *Historical Studies in the Physical Sciences*, **7**, 405–36.

Thomson, G. P. (1920a). *Applied Aerodynamics*. London: Hodder and Stoughton.

Thomson, G. P. (1920b). A note on the nature of the carriers of anode rays. *Proceedings of the Cambridge Philosophical Society*, **20**, 210–11.

Thomson, G. P. (1920c). The spectrum of hydrogen positive rays. *Philosophical Magazine*, **40**, 240–47.

Thomson, G. P. (1921). The application of anode rays to the investigation of isotopes. *Philosophical Magazine*, **42**, 857–67.

Thomson, G. P. (1922). The scattering of hydrogen positive rays, and the existence of a powerful field of force in the hydrogen molecule. *Proceedings of the Royal Society*, **102**, 197–209.

Thomson, G. P. (1925). A physical interpretation of Bohr's stationary states. *Philosophical Magazine*, **1**, 163–64.

Thomson, G. P. (1926a). The scattering of positive rays of hydrogen. *Philosophical Magazine*, **1**, 961–77.

Thomson, G. P. (1926b). An optical illusion due to contrast. *Proceedings of the Cambridge Philosophical Society*, **23**, 419–21.

Thomson, G. P. (1927). The diffraction of cathode rays by thin films of platinum. *Nature*, **120**, 802.

Thomson, G. P. (1928a). Experiments on the diffraction of cathode rays. *Proceedings of the Royal Society*, **117**, 600–9.

Thomson, G. P. (1928b). Experiments on the diffraction of cathode rays. II. *Proceedings of the Royal Society*, **119**, 651–63.

Thomson, G. P. (1928c). The disintegration of radium E from the point of view of wave mechanics. *Nature*, **121**, 615–16.

Thomson, G. P. (1928d). The waves of an electron. *Nature*, **122**, 279–82.

Thomson, G. P. (1929a). Experiments on the diffraction of cathode rays. III. *Proceedings of the Royal Society*, **125**, 352–70.

Thomson, G. P. (1929b). On the waves associated with β-rays, and the relation between free electrons and their waves. *Philosophical Magazine*, **7**, 405–17.

Thomson, G. P. (1929c). New discoveries about electrons. *The Listener*, **1**, 219–20.

Thomson, G. P. (1930). *The Wave Mechanics of Free Electrons*. New York & London: McGraw-Hill.

Thomson, G. P. (1938). Electronic waves. Nobel lecture, 7th June, available at: http://www.nobelprize.org/nobel_prizes/physics/laureates/1937/thomson-lecture.html.

Thomson, G. P. (1961). Early work in electron diffraction. *American Journal of Physics*, **29**, 821–5.

Thomson, G. P. (1963). Charles Galton Darwin. *Biographical Memoirs of Fellows of the Royal Society*, **9**, 69–85.

Thomson, G. P. (1964). *J. J. Thomson and the Cavendish Laboratory in his Day*. London: Nelson.

Thomson, G. P. (1966). *Autobiography*. Unpublished manuscript. Available in the archives of Trinity College Cambridge.

Thomson, G. P. (1968). The early history of electron diffraction. *Contemporary Physics*, **9**, 1–15.

Thomson, G. P. and Fraser, C. G. (1930). A camera for electron diffraction. *Proceedings of the Royal Society*, **128**, 641–8.

Thomson, G. P. and Reid, A. (1927). Diffraction of cathode rays by a thin film. *Nature*, **119**, 890.

Thomson, J. J. (1880). On Maxwell's theory of light. *Philosophical Magazine*, **9**, 284–91.

Thomson, J. J. (1881a). On the electric and magnetic effects produced by the motion of electrified bodies. *Philosophical Magazine*, **11**, 229–49.

Thomson, J. J. (1881b). On some electromagnetic experiments with open circuits. *Philosophical Magazine*, **12**, 49–60.

Thomson, J. J. (1883a). *On the Motion of Vortex Rings*. London: Macmillan & Co.

Thomson, J. J. (1883b). On the determination of the number of electrostatic units in the electromagnetic unit of electricity. *Philosophical Transactions of the Royal Society*, **174**, 707–21.

Thomson, J. J. (1883c). On a theory of the electric discharge in gases. *Philosophical Magazine*, **15**, 423–34.

Thomson, J. J. (1885). Some applications of dynamical principles to physical phenomena, I. *Philosophical Transactions*, **176**, 307–42.

Thomson, J. J. (1886). Some experiments on the electric discharge in a uniform electric field, with some theoretical considerations about the passage of electricity through gases. *Proceedings of the Cambridge Philosophical Society*, **5**, 391–409.

Thomson, J. J. (1887). Some applications of dynamical principles to physical phenomena, II. *Philosophical Transactions*, **178**, 471–526.

Thomson, J. J. (1888). *Applications of Dynamics to Physics and Chemistry*. London: Macmillan.

Thomson, J. J. (1889a). Specific inductive capacity of dielectrics when acted on by very rapidly alternating electric forces. *Proceedings of the Royal Society*, **46**, 292–5.

Thomson, J. J. (1889b). Note on the effect produced by conductors in the neighbourhood of a wire on the rate of propagation of electrical disturbances along it, with a determination of this rate. *Proceedings of the Royal Society*, **46**, 1–13.

Thomson, J. J. (1889c). The resistance of electrolytes to the passage of very rapidly alternating currents, with some investigations on the times of vibration of electrical systems. *Proceedings of the Royal Society*, **45**, 269–90.

Thomson, J. J. (1890a). On the passage of electricity through hot gases. I. *Philosophical Magazine*, **29**, 358–66.

Thomson, J. J. (1890b). Some experiments on the velocity of transmission of electric disturbances and their application to the theory of the striated discharge through gases. *Philosophical Magazine*, **30**, 129–40.

Thomson, J. J. (1891). On the illustration of the properties of the electric field by means of tubes of electrostatic induction. *Philosophical Magazine*, **31**, 149–71.

Thomson, J. J. (1893). *Notes on Recent Researches in Electricity and Magnetism*. Oxford: Clarendon Press.

Thomson, J. J. (1894a). The connection between chemical combination and the discharge of electricity through gases. In *Report of the British Association for the Advancement of Science, Oxford 1894*, pp. 482–93. London: J. Murray.

Thomson, J. J. (1894b). The electrolysis of steam. *Proceedings of the Royal Society*, **53**, 90–110.

Thomson, J. J. (1894c). On the velocity of the cathode-rays. *Philosophical Magazine*, **38**, 358–65.

Thomson, J. J. (1895a). The relation between the atom and the charge of electricity carried by it. *Philosophical Magazine*, **40**, 511–44.

Thomson, J. J. (1895b). On the electrolysis of gases. *Proceedings of the Royal Society*, **58**, 244–57.

Thomson, J. J. (1896a). The Röntgen rays. *Nature*, **53**, 391–2.

Thomson, J. J. (1896b). On the discharge of electricity produced by the Röntgen rays, and the effects produced by these rays on dielectrics through which they pass. *Proceedings of the Royal Society*, **59**, 274-6.

Thomson, J. J. (1896c). Presidential Address. Section A. *Report of the British Association for the Advancement of Science, 1896*, pp. 699-706. London: J. Murray.

Thomson, J. J. (1897a). On the cathode rays. *Proceedings of the Cambridge Philosophical Society*, **9**, 243-4.

Thomson, J. J. (1897b). *Cathode Rays*. London: Royal Institution.

Thomson, J. J. (1897c). On Cathode Rays. *Philosophical Magazine*, **44**, 293 - 316.

Thomson, J. J. (1898a). *The Discharge of Electricity Through Gases*. New York: Charles Scribner's Sons.

Thomson, J. J. (1898b). A theory on the connexion between cathode and Röntgen rays. *Philosophical Magazine*, **45**, 172-83.

Thomson, J. J. (1899a). On the theory of the conduction of electricity through gases by charged ions. *Philosophical Magazine*, **47**, 254-68.

Thomson, J. J. (1899b). On the masses of the ions in gases at low pressures. *Philosophical Magazine*, **48**, 547-67.

Thomson, J. J. (1900). The genesis of the ions in the discharge of electricity through gases. *Philosophical Magazine*, **50**, 278-83.

Thomson, J. J. (1901a). On the question as to whether or not there are any free charged ions produced during the combination of hydrogen and chlorine; and on the effect produced on the rate of the combination by the presence of such ion. *Proceedings of the Cambridge Philosophical Society*, **11**, 90-1.

Thomson, J. J. (1901b). On the theory of electric conduction through thin metallic films. *Proceedings of the Cambridge Philosophical Society*, **11**, 120-2.

Thomson, J. J. (1901c). The existence of bodies smaller than atoms. *Proceedings of the Royal Institution of Great Britain*, **16**, 574-86.

Thomson, J. J. (1902a). On some of the consequences of the emission of negatively electrified corpuscles by hot bodies. *Philosophical Magazine*, **4**, 253-62.

Thomson, J. J. (1902b). Experiments on induced-radioactivity in air, and on the electrical conductivity produced in gases when they pass through water. *Philosophical Magazine*, **4**, 352-67.

Thomson, J. J. (1902c). On the increase in the electrical conductivity of air produced by its passage through matter. *Proceedings of the Cambridge Philosophical Society*, **11**, 505.

Thomson, J. J. (1903a). *Conduction of Electricity through Gases*. Cambridge: Cambridge University Press.

Thomson, J. J. (1903b). The magnetic properties of systems of corpuscles describing circular orbits. *Philosophical Magazine*, **6**, 673-93.

Thomson, J. J. (1904a). *Electricity and Matter*. New York: Scribner.

Thomson, J. J. (1904b). On the structure of the atom: an investigation of the stability and periods of oscillation of a number of corpuscles arranged at equal intervals around the circumference of a circle; with application of the results to the theory of atomic structure. *Philosophical Magazine*, **7**, 237-65.

Thomson, J. J. (1905a). On the vibrations of atoms containing 4,5,6,7, and 8 corpuscles and on the effect of a magnetic field on such vibrations. *Proceedings of the Cambridge Philosophical Society*, **13**, 39–48.

Thomson, J. J. (1905b). The structure of the atom. *Proceedings of the Royal Institution of Great Britain*, **18**, 49–63.

Thomson, J. J. (1905c). On the emission of negative corpuscles by the alkali metals. *Philosophical Magazine*, **10**, 584–90.

Thomson, J. J. (1906a). A theory of widening of lines in the spectra. *Proceedings of the Cambridge Philosophical Society*, **13**, 318–21.

Thomson, J. J. (1906b). On the number of corpuscles in an atom. *Philosophical Magazine*, **11**, 769–81.

Thomson, J. J. (1907a). *The Corpuscular Theory of Matter*. London: Constable & Co.

Thomson, J. J. (1907b). The modern theory of electrical conductivity in metals. *Journal of the Institution of Electrical Engineers*, **38**, 455–65.

Thomson, J. J. (1907c). Rays of positive electricity. *Proceedings of the Royal Institution of Great Britain*, **18**, 1–16.

Thomson, J. J. (1907d). On rays of positive electricity. *Philosophical Magazine*, **13**, 561–75.

Thomson, J. J. (1907e). On the ionization of gases by ultra-violet light and on the evidence as to the structure of light afforded by its electrical effects. *Proceedings of the Cambridge Philosophical Society*, **14**, 417–24.

Thomson, J. J. (1907f). *On the Light Thrown by Recent Investigations on Electricity on the Relation between Matter and Ether. The Adamson Lecture Delivered at the University of Manchester on November 4*, Manchester: Manchester University Press.

Thomson, J. J. (1909a). Presidential address. *Report of the British Association for the Advancement of Science, 1909*, 3–29.

Thomson, J. J. (1909b). Positive electricity. *Philosophical Magazine*, **18**, 821–45.

Thomson, J. J. (1910a). On a theory of the structure of the electric field and its application to Röntgen radiation and to Light. *Philosophical Magazine*, **19**, 301–13.

Thomson, J. J. (1910b). On the theory of radiation. *Philosophical Magazine*, **20**, 238–47.

Thomson, J. J. (1910c). Rays of positive electricity. *Report of the British Association for the Advancement of Science, 1910*, 752–67.

Thomson, J. J. (1911). A new method of chemical analysis. *Proceedings of the Royal Institution of Great Britain*, **20**, 140–8.

Thomson, J. J. (1912a). The unit theory of light. *Proceedings of the Cambridge Philosophical Society*, **16**, 643–52.

Thomson, J. J. (1912b). Further experiments on positive rays. *Philosophical Magazine*, **24**, 209–53.

Thomson, J. J. (1912c). Multiply-charged atoms. *Philosophical Magazine*, **24**, 668–72.

Thomson, J. J. (1913a). On the structure of the atom. *Philosophical Magazine*, **26**, 792–99.

Thomson, J. J. (1913b). *Rays of Positive Electricity and their Application to Chemical Analyses*. London: Longmans, Green and Co. (second edition, 1921).

Thomson, J. J. (1914). The forces between atoms and chemical affinity. *Philosophical Magazine*, **27**, 757–89.

Thomson, J. J. (1917). Address of the president, sir J. J. Thomson, O. M., at the anniversary meeting, November 30, 1916. *Proceedings of the Royal Society*, **93**, 90–98.

Thomson, J. J. (1918). Address of the president, sir J. J. Thomson, O. M., at the anniversary meeting, November 30, 1917. *Proceedings of the Royal Society*, **94**, 182–90.

Thomson, J. J. (1919). On the origin of spectra and Planck's law. *Philosophical Magazine*, **37**, 419–66.

Thomson, J. J. (1920). On the scattering of light by unsymmetrical atoms and molecules. *Philosophical Magazine*, **40**, 393–413.

Thomson, J. J. (1923a). *The Electron in Chemistry: being five lectures delivered at the Franklin Institute, Philadelphia*. Philadelphia: Franklin Institute.

Thomson, J. J. (1923b). *Physics in Industry*, **1**. London: Institute of Physics.

Thomson, J. J. (1924). A suggestion as to the structure of light. *Philosophical Magazine*, **48**, 737–46.

Thomson, J. J. (1925). *The Structure of Light. The Fison Memorial Lecture*. Cambridge: Cambridge University Press.

Thomson, J. J. (1928a). Waves associated with moving electrons. *Philosophical Magazine*, **5**, 191–98.

Thomson, J. J. (1928b). Electronic waves and the electron. *Philosophical Magazine*, **6**, 1254–81.

Thomson, J. J. (1928c). *Beyond the Electron*. Cambridge: Cambridge University Press.

Thomson, J. J. (1930). *Tendencies of Recent Investigations in the Field of Physics, BBC Broadcast, 27th January 1930*. London: BBC.

Thomson, J. J. (1930/31). Atoms and Electrons. *Manchester Memoirs*, **75**, 77–93.

Thomson, J. J. (1936). *Recollections and Reflections*. London: G. Bell.

Thomson, J. J. (1939). Electronic waves. *Philosophical Magazine*, **27**, 1–33.

Thomson, J. J. and McClelland, J. A. (1896). On the leakage of electricity through dielectrics traversed by Röntgen rays. *Proceedings of the Cambridge Philosophical Society*, **9**, 126–40.

Thomson, J. J. and Newall, H. F. (1885). On the formation of vortex rings by drops falling into liquids, and some allied phenomena. *Proceedings of the Royal Society*, **39**, 417–36.

Thomson, J. J. and Rutherford, E. (1896). On the passage of electricity through gases exposed to Röntgen rays. *Philosophical Magazine*, **42**, 392–407.

Thomson, J. J. and Searle, G. (1890). A determination of 'v', the ratio of the electromagnetic unit of electricity to the electrostatic unit. *Philosophical Transactions of the Royal Society*, **181**, 583–621.

Thomson, J. J. and Threlfall, R. (1886). Some experiments on the production of ozone. *Proceedings of the Royal Society*, **40**, 340–2.

Thomson, W. (1867). On vortex atoms. *Proceedings of the Royal Society of Edinburgh*, **6**, 94–105.

Thomson, W. (1902). Aepinus atomized. *Philosophical Magazine*, **3**, 257–83.

Topper, D. R. (1971). Commitment to mechanism: J. J. Thomson, the early years. *Archive for the History of Exact Sciences*, **7**, 393–410.

Topper, D. R. (1980). To reason by means of images: J. J. Thomson and the mechanical picture of nature. *Annals of Science*, **37**, 31–57.

Turner, F. M. (1993). *Contesting Cultural Authority: Essays in Victorian Intellectual Life*. Cambridge: Cambridge University Press.

Turpin, B. (1980). *The Discovery of the Electron: the Evolution of a Scientific Concept*. Unpublished Ph.D. dissertation. University of Notre Dame.

Warwick, A. (2003a). *Masters of Theory. Cambridge and the Rise of Mathematical Physics*. Chicago: Chicago University Press.

Warwick, A. (2003b). 'That universal aethereal plenum': Joseph Larmor's natural history of physics. In *From Newton to Hawking. A history of Cambridge University's Lucasian Professors of Mathematics*, eds. K. C. Knox and R. Noakes. Cambridge: Cambridge University Press.

Wheaton, B. R. (1983). *The Tiger and the Shark. Empirical Roots of Wave-Particle Duality*. Cambridge: Cambridge University Press.

Whittaker, E. T. (1951). *A History of the Theories of Aether and Electricity. 2 vols*. London: Nelson.

Williams, P. (1990). Passing the torch: Whewell's philosophy and the principles of English university education. In *William Whewell: A Composite Portrait*, eds. M. Fisch and S. Schaffer, pp. 117–47. Oxford: Oxford University Press.

Wien, W. (1898). Die electrostatische und magnetische Ablenkung der Kanalstrahlen. *Berlin Physikalische Gesellschaft Verhandlungen*, **17**, 10–2.

Wilson, D. (1983). *Rutherford. Simple Genius*. London: Hodder and Stoughton.

Index

183

Printed in the United States
By Bookmasters